W9-ADY-941

612.405 81882
B52

DATE DUE			
Feb27'73			
Apr5'76			
Nov5 78			
Apr26 79			
Nov2 79			
Apr24 '81			
Nov24 '82			

DISCARDED

Biochemical Responses to Environmental Stress

Biochemical Responses to Environmental Stress

Proceedings of a Symposium sponsored by the Divisions of Water, Air, and Waste Chemistry, Microbial Chemistry and Technology, and Biological Chemistry of the American Chemical Society, held in Chicago, Illinois, September 14-15, 1970.

Edited by I. A. Bernstein

Professor of Environmental and
Industrial Health and of Biological Chemistry
The University of Michigan
Ann Arbor, Michigan

Plenum Press · New York-London · 1971

CARL A. RUDISILL LIBRARY
LENOIR RHYNE COLLEGE

612.405
B52
81882
Jan. 1973

Library of Congress Catalog Card Number 79-151618

SBN 306-30531-3

© 1971 Plenum Press, New York
A Division of Plenum Publishing Corporation
227 West 17th Street, New York, N.Y. 10011

United Kingdom edition published by Plenum Press, London
A Division of Plenum Publishing Company, Ltd.
Davis House (4th Floor), 8 Scrubs Lane,
Harlesden, NW10 6SE, England

All rights reserved

No part of this publication may be reproduced in any form
without written permission from the publisher

Printed in the United States of America

PREFACE

In the world outside the laboratory, life goes on in a changing rather than in a constant environment and organisms must continually accommodate to changes in temperature, light, humidity, nutrition, etc. Since studies of the enzymatic process, *in vitro*, indicate that, in general, biological catalysis can proceed only over limited ranges of temperature, pH, substrate concentration, etc., it seems reasonable to assume that biological systems have an ability to maintain a relatively constant internal milieu in the face of drastic external environmental change.

This concept, as applied particularly to the mammal, was enunciated by Bernard (1878) in the latter part of the last century. Cannon (1939) designated the phenomenon as *homeostasis* stating (cf Potter, 1970) that "in an open system such as our bodies represent, compounded of unstable material and subjected continually to disturbing conditions, constancy is in itself evidence that agencies are acting or are ready to act, to maintain this constancy." He further proposed that "if a state remains steady, it does so because any tendency towards change is automatically met by increased effectiveness of the factor or factors which resist the change."

Considerable evidence (cf Prosser, 1958) suggests that homeostasis is a general phenomenon which applies to all living things and at all levels of biological complexity. Survival in the face of environmental stress would seem to depend upon the ability of the organism to respond by appropriate biochemical modulations so as to maintain homeostasis. Obviously, a stress requiring an adjustment in excess of a living system's ability to accommodate, will be toxic or will result in disease.

This symposium focuses on specific examples of environmental accommodation in microbial and animal systems with particular emphasis on the molecular mechanisms involved. The studies chosen for presentation are by no means all the available information on the subject, nor is the present organization of the material the first to appear in the literature. Investigators have, in reality, been studying environmental effects on biological systems for decades. It now appears worthwhile to bring their work into perspective in context of society's demand for control of further environmental pollution. An understanding of the cell's molecular mechanisms of response to environmental change and the degree of stress to which the cell can accommodate through these mechanisms, appears essential to a rational definition of environmental quality standards.

Although it is already clear that organisms have an extensive ability to accommodate quickly, it is still necessary to determine the degree to which this is possible and the degree to which a first response affects the system's ability to accommodate to a second stress. The increasingly rapid alteration of the environment resulting from man's technological development makes it important to evaluate this latter proposition. Our ability to accommodate to a particular stress may be seriously limited by prior responses to other environmental insults.

The homeostatic capabilities of biological systems should not be taken as an excuse for allowing environmental contamination to continue unabated. Rather, these abilities should be taken into account in setting priorities and limits for control of environmental pollution.

The Editor has tried to refrain from significantly influencing either the context or the rhetoric of the presentations in the hope of providing the reader with a more comprehensive view of the subject. He does, however, assume complete responsibility for the choice of contributors believing that the papers do provide a balanced - even though incomplete - review of the information relevant to the title of the symposium.

Thanks are due the authors for their cooperation in expeditiously providing typed manuscripts. The Editor is especially grateful for the assistance of Mrs. Barbara Gooding in preparing the material for submission to the Publisher. Permission from the following journals to reprint specific figures in the text is acknowledged with thanks: Proceedings of the National Academy of Sciences (US), Science, Biophysical Journal, Nature, Cancer Research, Journal of Membrane Biology, American Journal of Physiology and Federation Proceedings.

Ann Arbor, Michigan
November 27, 1970

I. A. Bernstein

REFERENCES

Bernard, C. 1878. Lecons sur les phénomènes la vie. Baillière, Paris.

Canon, W.B. 1939. The wisdom of the body. Norton, New York.

Potter, V.R. 1970. Intracellular responses to environmental change: the quest for optimum environment. Environ. Res. 3: 176-186.

Prosser, C.L. The nature of physiological adaptation. p. 167-180. In C.L. Prosser [Ed], Physiological adaptation. Am. Physiol. Soc., Washington.

CONTRIBUTORS TO THIS VOLUME

JERE M. BAUER
Department of Internal Medicine
The University of Michigan Medical Center
Ann Arbor, Michigan 48104

I. A. BERNSTEIN
Departments of Environmental and Industrial Health and Biological Chemistry
The University of Michigan Medical Center
Ann Arbor, Michigan 48104

ROBERT E. BEYER
Laboratory of Chemical Biology
Department of Zoology
The University of Michigan
Ann Arbor, Michigan 48104

THOMAS D. BROCK
Department of Microbiology
Indiana University
Bloomington, Indiana 47401

FRED R. BUTCHER
McArdle Laboratory
University of Wisconsin
Madison, Wisconsin 53706

J. E. DONNELLAN, JR.
Biology Division
Oak Ridge National Laboratory
Oak Ridge, Tennessee 37830

M. ESFAHANI
Biochemistry Department
Duke University Medical Center
Durham, North Carolina 27706

DAVID KUPFER
Lederle Laboratories
Division of American Cyanamid Company
Pearl River, New York 10965

R. P. MORTLOCK
Department of Microbiology
University of Massachusetts
Amherst, Massachusetts 01002

VAN R. POTTER
McArdle Laboratory
University of Wisconsin
Madison, Wisconsin 53706

ROBERT D. REYNOLDS
McArdle Laboratory
University of Wisconsin
Madison, Wisconsin 53706

DAVID F. SCOTT
McArdle Laboratory
University of Wisconsin
Madison, Wisconsin 53706

R. S. STAFFORD
Biology Division
Oak Ridge National Laboratory
Oak Ridge, Tennessee 37830

SALIH J. WAKIL
Biochemistry Department
Duke University Medical Center
Durham, North Carolina 27706

W. A. WOOD
Department of Biochemistry
Michigan State University
East Lansing, Michigan 48823

CHUNG WU
Departments of Internal Medicine
and Biological Chemistry
The University of Michigan Medical Center
Ann Arbor, Michigan 48104

CONTENTS

ABBREVIATIONS

AMP   Adenosine monophosphate

ADP   Adenosine diphosphate

ATP   Adenosine triphosphate

CoA   Coenzyme A

CoQ   Coenzyme Q

DNA   Deoxyribonucleic acid

FeNH   "Non-heme iron"

FP   Flavoprotein

NAD   Nicotinamide adenine nucleotide (or diphosphopyridine nucleotide = DPN)

NADH   Reduced nicotinamide adenine nucleotide

Pi   Inorganic phosphate

RNA   Ribonucleic acid

$\widehat{TT}$ Thymine dimer

$\widehat{UT}$ Uracil-thymine dimer

GENETIC AND ENZYMATIC MECHANISMS FOR THE ACCOMMODATION TO NOVEL SUBSTRATES BY *AEROBACTER AEROGENES*

R. P. Mortlock and W. A. Wood

Department of Microbiology, University of Massachusetts

Amherst, Mass. 01002 and Department of Biochemistry

Michigan State University, East Lansing, Mich. 48823

Modern technology is presenting to our environment a number of unnatural compounds. The accumulation of these strange or "xeno-chemicals", some of which are quite toxic, places a stress upon the natural environment which threatens to bring about drastic changes. Future attempts to deal with this problem must capitalize on all potential agents with abilities to act upon xenochemicals. In particular, the bacteria have evolved the metabolic capacity to utilize a great variety of organic compounds as sources of nutrients and energy. In this connection, many of the new xenochemicals accumulating in our environment are a potential source of carbon and energy for the growth of microorganisms and, hence, present a challenge to the organisms' versatility to derive the enzymatic ability to utilize these new compounds.

Microorganisms currently residing in the biosphere are there because of their ability to metabolize compounds found in the environment. There is no reason to believe that the ability to attack materials not found in nature would evolve or persist in these organisms. Yet it is easy to document the fact that a wide variety of xenocompounds is attacked by a number of microbial species. What are the mechanisms by which organisms are able to react to this type of stress situation so that the organisms survive? If we could know those processes, they might be exploited to combat the accumulation of xeno compounds in the environment.

During the course of evolution, increased diversity in the biosynthetic abilities of organisms must have continually presented new chemical structures to the natural environment. New organisms then appeared with an ability to use such compounds as carbon and energy sources. With the unnatural, man-made chemicals, the difference in structure from pre-existent natural compounds might be so great as to require a number of mutational events to permit efficient utilization. If several separate mutations were required to permit rapid degradation of some xenochemical, the probability of all the mutations occurring simultaneously in a single organism would be extremely low. Whereas such rare events might occur in the course of biochemical evolution, there is no such time available for selection if we wish to use microorganisms to scrub our present environment. However, if we could predict which mutations were needed and select for these by sophisticated laboratory manipulations we could program the development of cells with the correct genetic information so as to enable them to utilize xeno compounds.

THE XENOPENTOSES AND XENOPENTITOLS

As a model system for the study of bacteria on xenochemicals we have been studying the metabolism of certain of the 5-carbon sugars. The aldopentoses exist in eight possible structures; L- and D-ribose, L- and D-xylose, L- and D-arabinose and L- and D-lyxose. Three of these eight structures, L-arabinose, D-ribose and D-xylose can be considered as naturally occurring. Many bacteria, including the coliform bacteria, have developed inducible enzyme pathways to utilize these sugars. The other five aldopentose structures are not common in nature and can be considered to be xenocarbohydrates. Only a limited number of organisms can utilize these, and events other than enzyme induction must occur before growth is possible. Similarly, for the four possible pentitols, ribitol and D-arabitol are naturally occurring while L-arabitol and xylitol are xenopentitols. Coliform bacteria of the _Aerobacter-Klebsiella_-type have inducible enzyme pathways for the utilization of the two naturally occurring pentitols, ribitol and D-arabitol (Mortlock and Wood, 1964a). Thus among this group of twelve C-5 structures, seven are considered to be xenocarbohydrates.

Although a bacterium such as _Aerobacter aerogenes_ possesses the genetic information to permit it to utilize the five naturally occurring sugars as growth substrates normal cells will not utilize or degrade any of these unnatural sugars. If about 10^8 cells are suspended in a growth medium containing one of the unnatural sugars as the sole carbon and energy source, then after an incubation period of one or more days, growth will occur accompanied by disappearance of the xeno sugar. From this culture mutants can be isolated which have apparently acquired a new enzymatic ability and are now capable of utilization of the xeno sugar. In such a manner, mutants can be readily selected which utilize each of these xeno-pentoses and pentitols with the possible exception of L-ribose, where the lack of sufficient quantity of pure L-ribose has prevented careful investigation (Mortlock and Wood, 1964b).

ENZYMATIC PATHWAYS FOR DEGRADATION

In order to establish the events that confer an ability to attack xenocarbohydrates, it has been necessary to establish their routes of utilization. The metabolic pathways for their dissimilation are shown in Figure 1. (From Mortlock, et al., 1965)

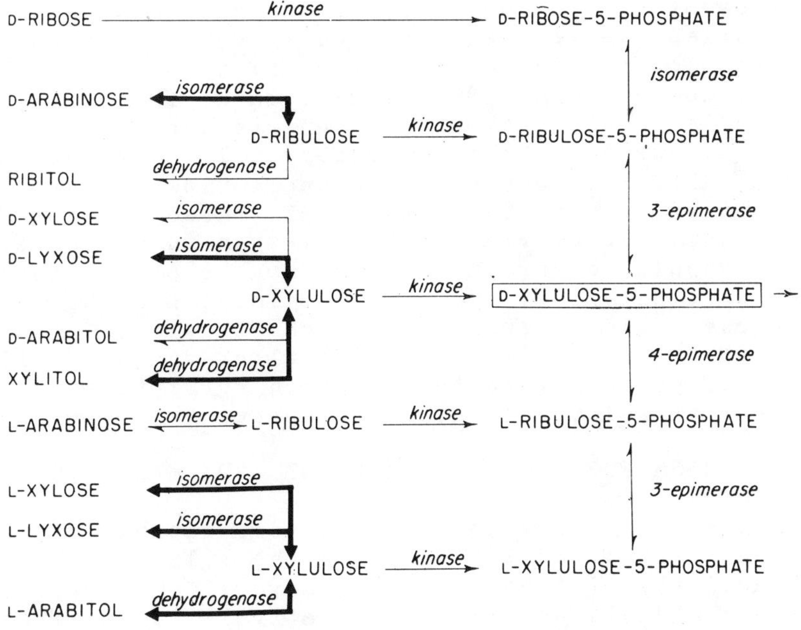

Fig. 1. Pathways for degradation of pentoses and pentitols by *Aerobacter aerogenes*.

With the exception of D-ribose, these aldopentoses are degraded by similar metabolic strategies. The first reaction is isomerization to the corresponding 2-ketopentose, followed by phosphorylation to the ketopentose 5-phosphate. Epimerization reactions convert the various ketopentose 5-phosphates to D-xylulose 5-phosphate, the common intermediate in the degradation of these compounds. Similarly the pentitols are utilized by a common strategy. The pentitols are oxidized to the 2-ketopentose and then follow the same route to D-xylulose 5-phosphate (Wood, 1966). The same metabolic pathways are followed by mutants utilizing one of the unnatural pentoses. The mutants which are capable of growth on the unnatural substrates have obtained the enzymatic activity necessary to convert (heavy arrows) the new substrate to a structure which is a metabolic intermediate in the degradation of a naturally occurring carbohydrate. From this point on, degradation can be by the normal enzymatic activity of the cell (light arrows).

What kind of mutational event can permit a cell to acquire such a new type of enzymatic ability? It had been assumed that this normally would result from change within the structural gene for an enzyme in such a manner as to alter the substrate affinity or catalytic properties of the enzyme. If a mutational change in a single nucleotide were all that were required to obtain the correct change in the structure of the enzyme, then this mutation could occur at a measurable frequency, given the large populations of bacteria that it is possible to survey in the laboratory. If, however, a number of different and specific nucleotide changes were required, the probability of all these happening simultaneously or over a short time span in a single organism should be rare. Mutants capable of growth on D-arabinose, D-lyxose, xylitol, L-xylose, L-arabitol or L-lyxose could be selected reproducibly from populations of from 10^7 to 10^8 cells (Mortlock and Wood, 1964b). For each of these sugars the mutation permitting growth must occur at relatively high frequency.

THE MECHANISM FOR GROWTH ON XYLITOL AND L-ARABITOL

In the study of mutants which were capable of growth on xylitol and in measuring the xylitol dehydrogenase activity of these mutants, several observations were made: 1) In addition to relatively low activity for the oxidation of xylitol to D-xylulose, such mutants contained very high levels of activity for the oxidation of ribitol to D-ribulose; that is, high levels of ribitol dehydrogenase; 2) During the course of purification, these two dehydrogenase activities could not be separated; 3) When mutants deficient in ribitol dehydrogenase were used as the parent strain, there was great difficulty in selecting mutants for growth on xylitol. These results suggested that the two dehydrogenase activities might be catalyzed by the same enzyme. This prediction was confirmed by further study (Mortlock et al., 1965). Ribitol dehydrogenase, an enzyme which had presumably evolved for the ribitol pathway, possessed the gratuitous ability to oxidize xylitol and therefore the degradation of xylitol did not require a new or altered enzyme. The enzymes of the ribitol pathway, however, were normally regulated and inducible. If ribitol was present in the medium, the cells synthesized ribitol dehydrogenase and the second enzyme for ribitol degradation, D-ribulokinase. As a xenocarbohydrate, xylitol was not recognized as an inducer of these enzymes. Thus actual growth on xylitol required a regulatory mutation, to permit the synthesis of the enzymes of the ribitol pathway in the absence of the inducer. Mutants selected for growth on xylitol were constitutive for the enzymes of the ribitol pathway; that is, the enzymes for ribitol utilization were always synthesized whether ribitol was present or not. During the growth of such mutants on xylitol the ribitol dehydrogenase catalyzed the oxidation of xylitol to D-xylulose. It was also observed that purified ribitol dehydrogenase was capable of catalyzing the oxidation of another xenopentitol, L-arabitol, to L-xylulose. Growth on L-arabitol also was found to result from a regulatory mutation which permitted the constitutive synthesis of ribitol dehydrogenase (Mortlock et al., 1965). Table 1 compares the range of ribitol dehydrogenase activities found in a variety of xylitol-positive or L-arabitol-positive mutants with the activity present in the wild-type parent strain.

Ribitol dehydrogenase has been purified and crystalized from such constitutive mutants and it has been shown that the enzyme will catalyze the oxidation of xylitol to D-xylulose or L-arabitol to L-xylulose. The relative Vmax for pentitol oxidation was 10/1/1 for ribitol, xylitol and L-arabitol, respectively. The constitutive enzyme could not be distinguished in its physical or immunological properties from the enzyme purified from ribitol-induced cells (Mortlock et al., 1965).

TABLE 1

RIBITOL DEHYDROGENASE ACTIVITY OF CELL-FREE EXTRACTS

Strain	Growth Substrate	Specific Activity* (μmoles/min./mg protein)
Wild-type PRL-R3	Ribitol	1.76 - 2.9
	CH**	0 - 0.2
Xylitol Positive	CH	2.6 - 6.3
L-Arabitol	CH	3.9 - 5.8

*Measured spectrophotometrically in the direction of D-ribulose reduction. Activities presented are high and low values from numerous experiments with different mutants.
**Casein Hydrolysate

Xylitol-positive mutants which had lost ribitol dehydrogenase activity simultaneously lost the ability to grow with either ribitol or xylitol as substrate. Furthermore, with a different organism, a strain of Klebsiella which can be transduced, the mutation which permits growth on xylitol appears to be located adjacent to the ribitol dehydrogenase and D-ribulokinase structural genes (Charnetzky and Mortlock, 1970).

Thus growth on both of these rare pentitols, xylitol and L-arabitol, initially resulted not from an alteration in a structural gene giving rise to a new enzyme, but from a mutation in a regulatory mechanism allowing the cell to synthesize an enzyme in the absence of its normal inducer. The ribitol dehydrogenase synthesized by the constitutive mutants apparently was unaltered from the induced enzyme and could catalyze the conversion of the rare pentitols to structures which could be further degraded by the existing enzymes of the cell.

The genetics of the regulation of the ribitol pathway are still under study. If the regulation of this pathway was under the conventional negative control then any mutational event which resulted in the production of inactive repressor should also result in constitutive synthesis of the enzymes of the pathway. Under the fail-safe system of negative control, the loss of function of a gene -- in this case the regulator gene of the ribitol pathway -- would enable the mutant to grow on xylitol. Mutations for loss of a genetic function can occur at relatively high frequency.

Once growth was established on some xenocarbohydrate by this type of mechanism, then gradual selection could occur for less frequent mutations to improve the growth rate on the new substrate. In the case of xylitol, either a modification in the dehydrogenase structural gene to permit more efficient catalysis of xylitol oxidation or an increase in the amount of ribitol dehydrogenase per cell should result in more rapid growth with xylitol as the substrate. Both of these events have been observed. Wu, et al., (1968), after continuous cultivation of cells on xylitol, have reported the isolation of a second mutant possessing an apparent alteration in the dehydrogenase structural gene. With our strain of *Aerobacter*, after continuous cultivation in a chemostat on xylitol a second mutation was obtained which increased the growth rate on xylitol from 0.26 to 0.5 generations per hour. No alteration in the activity of the enzyme for xylitol or ribitol could be observed but the constitutive dehydrogenase level found in crude extracts of these new mutant cells was elevated 4-fold, from 7.5 to 31 units per mg protein (Bisson, 1968).

THE MECHANISM FOR GROWTH ON D-ARABINOSE

Aerobacter aerogenes

It was of interest to explore the other unnatural pentoses to determine if a similar mechanism was used to establish growth; that is, a mutation of the control mechanism for an existing enzyme which could then catalyze the initial step in degradation of a xenopentose. Investigation showed that growth of *Aerobacter* on D-arabinose also resulted from a regulatory mutation. In this case the required catalytic activity for the isomerization of D-arabinose to D-ribulose was supplied by an enzyme of the L-fucose catabolic pathway, L-fucose isomerase as illustrated in Figure 2.

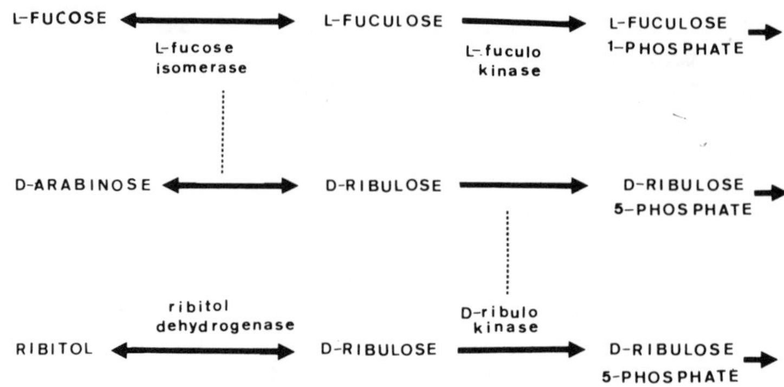

Fig. 2. Origin of enzymes used for D-arabinose degradation by *Aerobacter aerogenes*.

Although this enzyme was normally utilized for L-fucose catabolism it possessed the ability to isomerize D-arabinose to D-ribulose. The enzyme was induced in wild-type cells upon incubation with L-fucose but not D-arabinose and mutants selected for growth on D-arabinose were constitutive for L-fucose isomerase (Camyre and Mortlock, 1965; Oliver, 1969). The D-ribulose produced by the isomerization of D-arabinose induced both of the enzymes of the ribitol pathway, ribitol dehydrogenase and D-ribulokinase (Bisson, et al., 1968). In this way, D-arabinose was converted to D-ribulose 5-phosphate which was on the main pathway for pentose degradation.

Cells which were not constitutive for the fucose enzymes would grow on D-arabinose under either of two conditions. In the first, wild-type cells were grown on L-fucose, washed, and transferred to D-arabinose. The pre-induced L-fucose isomerase permitted isomerization of D-arabinose and, after a suitable lag to permit the induction of D-ribulokinase, growth was initiated. Growth in this case was linear since the inducer of the isomerase was no longer present and the isomerase activity per ml of medium was constant. In the second method a mutant with a block in the L-fucose pathway after the isomerase step was obtained from wild-type cells. When such mutant cells were incubated in a D-arabinose medium containing a small amount of L-fucose, the fucose continuously induced the isomerase and growth using D-arabinose was exponential and comparable to that obtained with the isomerase constitutive mutants (Oliver, 1969).

TABLE 2

D-ARABINOSE ISOMERASE ACTIVITY OF CELL-FREE EXTRACTS

Strain	Growth Substrate	Isomerase activity (umoles/min/mg protein)		Ratio A/F Activity
		D-Arabinose	L-Fucose	
Wild-Type PRL-R3	L-Fucose CH*	1.4 0**	2.35 0	0.6/1
502	CH	0.97	1.38	0.7/1
531	CH	3.16	1.9	1.65/1
534	CH L-Fucose	0.19 4.63	0.15 2.8	- 1.6/1

*Casein hydrolysate
**Activity not detectable, less than 0.05

Table 2 shows the isomerase activities in crude cell-free extracts of wild-type cells induced with L-fucose, as compared with the constitutive activity of one of the D-arabinose-positive mutants, strain 502. The isomerase has been purified from induced and constitutive cells and the ratio of arabinose to fucose activity remained constant throughout the purification technique (Oliver and Mortlock, 1968). After several years in stock culture with D-arabinose as the substrate, mutant strain 531 was isolated possessing an apparent alteration in the isomerase catalytic activity. Whereas the normal D-arabinose/L-fucose ratio of activity was 0.6/1, this new mutant possessed a ratio of 1.6/1. When purified, the new isomerase displayed an alteration in Vmax and a decrease in the Km for D-arabinose. Perhaps this is a second stage mutation and the beginning of the evolution of a true D-arabinose isomerase. From this latter, mutant strain 534 was obtained with regulation of the enzyme partially restored. This strain grew very slowly using D-arabinose but would still grow normally using fucose and now L-fucose induced the modified ratio of isomerase activity.

L-fucose isomerase also catalyzes the isomerization of the xenopentose, L-xylose, to L-xylulose and mutants selected for growth on L-xylose were found to be constitutive for this enzyme. Thus for at least four of the unnatural sugars, the initial mutational event permitting growth of *Aerobacter* is a regulatory mutation. For xylitol and L-arabitol, the regulatory mutation resulted in the constitutive synthesis of ribitol dehydrogenase. For D-arabinose, L-xylose and possibly L-lyxose, growth resulted from a mutation to constitutive synthesis of the enzymes of the L-fucose catabolic pathway with L-fucose isomerase supplying the necessary isomerase activity (Figure 1). An exception to this pattern may exist for growth on D-lyxose, which has been investigated by Anderson and Allison (1965). These latter workers purified D-lyxose isomerase and could not identify this enzyme activity as a secondary activity of a more common enzyme.

Escherichia coli

Mutants of *Escherichia coli* strain K-12 also could be selected for growth on D-arabinose and the mutation appeared to be a regulatory mutation affecting the L-fucose catabolic pathway. However, as shown in Table 3, in this case the mutation permitted L-fucose isomerase to be induced upon incubation of the mutant strain in the D-arabinose medium. This was different than in *Aerobacter* where constitutivity resulted from mutation. For both organisms, the L-fucose isomerase catalyzed the isomerization of D-arabinose to D-ribulose (Green

TABLE 3

INDUCTION OF L-FUCOSE ISOMERASE IN *ESCHERICHIA COLI* K-12, AND A D-ARABINOSE POSITIVE MUTANT

Strain	Carbohydrate in Growth Medium*	Isomerase Specific Activity (nanomoles/min/mg protein)	
		D-Arabinose	L-Fucose
Wild-Type K-12	None	0 - 2	0 - 7
	D-Arabinose	0 - 2	0 - 7
	L-Fucose	87	249
D-Arabinose Positive Mutant	None	0 - 2	0 - 8
	D-Arabinose	74	165
	L-Fucose	72	214

and Cohen, 1956; LeBlanc and Mortlock, 1970). Since Escherichia does not possess a ribitol degrading pathway it does not have a natural D-ribulokinase to carry out the second step of D-arabinose degradation. As shown in Figure 3, the organism used additional enzymes of the L-fucose pathway to complete the degradation of D-arabinose.

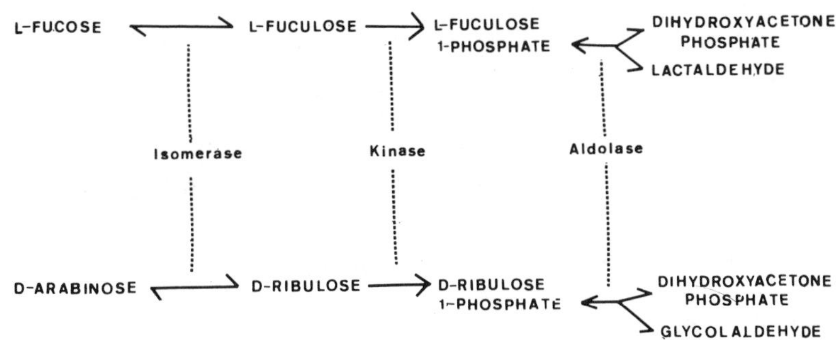

Figure 3. Origin of enzymes for the degradation of D-arabinose by Escherichia coli K-12.

L-Fuculokinase phosphorylated D-ribulose to D-ribulose 1-phosphate and fuculose 1-phosphate aldolase cleaved the ribulose 1-phosphate to dihydroxyacetone phosphate and glycolaldehyde (Ghalambor and Heath, 1962;LeBlanc and Mortlock, 1970). The glycolaldehyde was oxidized to glycolic acid which accumulated in the medium while the D-arabinose was in excess. All three of these enzymes were induced when D-arabinose-positive mutants were incubated with D-arabinose. Mutants which had lost the activity of any of these three enzymes lost the ability to grow on either L-fucose or D-arabinose.

If E. coli K-12 was capable of phosphorylating D-ribulose at the 5-carbon position, this would be a more direct route for degradation and should result in faster growth on D-arabinose. In fact, the organism possesses the ability to synthesize an enzyme to catalyze this reaction. As illustrated in Figure 4, the enzyme is the L-ribulokinase of the L-arabinose catabolic pathway which, in addition to its normal substrate L-ribulose, is known to catalyze the phosphorylation of D-ribulose, and at the five carbon position (Lee and Bendet, 1966). This enzyme is induced in wild-type cells by L-arabinose. When D-arabinose-positive cells were used as parent strain to select mutants which were constitutive for the L-arabinose enzymes, the growth rate on D-arabinose increased from 0.32 to 0.71 generations per hour and glycolic acid no longer accumulated in the medium. Another method of programming the cells to contain L-ribulokinase, employed a mutant blocked in the first enzyme of the L-arabinose pathway, L-arabinose isomerase.

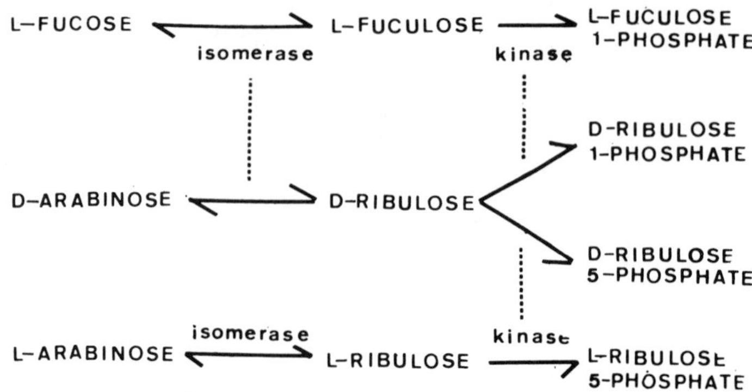

Figure 4. Alternate routes for D-arabinose degradation by *E. coli* K-12.

The mutant could not use L-arabinose as a growth substrate but small amounts of L-arabinose in the medium caused the continuous induction of the L-ribulokinase. When a D-arabinose-positive, L-arabinose isomerase-negative, mutant was grown on D-arabinose, the growth rate increased from 0.31 to 0.83 generations per hour following the addition of 0.005 % L-arabinose to the medium (LeBlanc, 1970).

It should be pointed out that attempts to involve enzymes in pathways for which they have not evolved can lead to special problems. In the case of L-mannose, studied by Mayo and Anderson, (1969), the wild-type organism was capable of synthesizing the required enzyme activity to degrade L-mannose through several reactions. A mutation was still required to permit growth on L-mannose, apparently to overcome the toxicity of a product of these reactions.

Under certain conditions there appears to be an advantage for microorganisms to possess enzymes which are flexible in their ability to metabolize different substrates. Perhaps early in the course of evolution primitive enzymes were capable of slowly catalyzing their reaction for a variety of substrates, giving a great versatility to the metabolic capabilities of the organism. Since natural selection has presumably operated for organisms with rapid catabolic pathways, specialization may be the price paid to obtain rapid growth rates.

ACKNOWLEDGEMENTS

This work was supported by PHS Research Grant No. 06848 from the Institute of Allergy and Infectious Diseases and National Science Foundation Grants G-6465 and G-21264.

REFERENCES

Anderson, R.L. and D.P. Allison. 1965. Purification and characterization of D-lyxose isomerase. J. Biol. Chem. 240:2367-2372.

Bisson, T.M. 1968. Regulation of five-carbon sugar metabolism in *Aerobacter aerogenes*: Ph.D. Dissertation, University of Massachusetts, Amherst, Mass.

Bisson, T.M., E.J. Oliver and R.P. Mortlock. 1968. Regulation of pentitol metabolism by *Aerobacter aerogenes*. II. Induction of the ribitol pathway. J. Bacteriol. 95:932-936.

Camyre, K.P. and R.P. Mortlock. 1965. Growth of *Aerobacter aerogenes* on D-arabinose and L-xylose. J. Bacteriol. 90:1157-1158.

Charnetzky, W.T. and R.P. Mortlock. 1970. Regulation of the ribitol catabolic pathway in *Klebsiella aerogenes*. Bacteriol. Proc. 137.

Ghalambor, M.A. and E.C. Heath. 1962. The metabolism of L-fucose. II. The enzymatic cleavage of L-fuculose 1-phosphate. J. Biol. Chem. 237:2427-2433.

Green, M. and S.S. Cohen. 1956. Enzymatic conversion of L-fucose to L-fuculose. J. Biol. Chem. 219: 557-568.

LeBlanc, D.J. 1970. Pathways of D-arabinose degradation among coliforms. Ph.D. Dissertation, University of Massachusetts, Amherst, Mass.

LeBlanc, D.J. and R.P. Mortlock. 1970. Pathways of D-arabinose degradation among coliforms. Bacteriol. Proc. 146.

Lee, N. and Z. Bendet. 1967. Crystalline L-ribulokinase from *Escherichia coli*. J. Biol. Chem. 242:2043-2050.

Mayo, J.W. and R.L. Anderson. 1969. Basis for the mutational acquisition of the ability of *Aerobacter aerogenes* to grow on L-mannose. J. Bacteriol. 100: 948-955.

Mortlock, R.P. and W.A. Wood. 1964(a). Metabolism of pentoses and pentitols by *Aerobacter aerogenes*. II. Mechanism of acquisition of kinase, isomerase and dehydrogenase activity. J. Bacteriol. 88:845-849.

Mortlock, R.P. and W.A. Wood. 1964(b). Metabolism of pentoses and pentitols by *Aerobacter aerogenes*. I. Demonstration of pentose isomerase, pentulokinase and pentitol dehydrogenase enzyme families. J. Bacteriol. 88:838-844.

Mortlock, R.P., D.D. Fossitt and W.A. Wood. 1965. A basis for utilization of unnatural pentoses and pentitols by *Aerobacter aerogenes*. Proc. Nat. Acad. Sci. (US) 54:572-579.

Oliver, E.J. 1969. Growth of *Aerobacter aerogenes* on D-arabinose. Ph.D. Dissertation, University of Massachusetts, Amherst, Mass.

Oliver, E.J. and R.P. Mortlock. 1968. Growth of *Aerobacter aerogenes* on D-arabinose. Bacteriol. Proc. 123.

Wood, W.A. 1966. Carbohydrate Metabolism. Ann. Rev. Biochem. 35:521-558.

Wu, T.T., E.C.C. Lin and S. Tanaka. 1968. Mutants of *Aerobacter aerogenes* capable of utilizing xylitol as a novel carbon. J. Bacteriol. 96:447-456.

THE RESPONSE OF *ESCHERICHIA COLI* TO FATTY ACID SUPPLEMENTS AND THE REGULATION OF MEMBRANE LIPID SYNTHESIS

Salih J. Wakil and M. Esfahani

Biochemistry Department, Duke University

Medical Center, Durham, N.C. 27706

Despite many years of research, the exact nature of the molecular organization of the biological membrane is still a matter of much discussion and speculation. Several models have been proposed for membrane structure (Danielli and Davson, 1935; Robertson, 1966; Benson, 1968; Green et al, 1967), each of which is based primarily on what is assumed for its function, and none of which has met a universal acceptability. However, what is generally agreed on is that the biomembrane consists of proteins, lipids, and water, and that lipid-protein interactions determine membrane function. A possible approach to the understanding of the structural requirements for the lipids and proteins which are essential for the expression of membrane function is alteration of these constituents and investigation into the effect of such alterations on membrane properties. To this purpose, alterations of the lipid component of the membrane have sounded promising (Silbert et al, 1968; McElhaney and Tourtellotte, 1969; Steim et al, 1969).

For several reasons, bacterial systems offer an advantage over others for studies relating membrane structure and function. Genetic regulations, cellular division, and mechanism of permeation are best understood in these organisms. Furthermore, variants of the population can be isolated with relative ease, as auxotrophs or conditional mutants with lesions in certain biological processes involved in membrane genesis and/or maintenance. These mutants can then be placed under conditions which induce desired alterations in the cell membrane. The effect of such alterations on various biological functions of the cell can be investigated.

Bacterial membrane is the site of oxidative phosphorylation (Marr, 1960), cell envelope synthesis (Salton, 1967), "permeases"

(Fox and Kennedy, 1965; Kundig and Roseman, 1969; Milner and Kaback, 1970), and probably initiation of chromosome replication (Jacob et al, 1963; Ganeson and Lederberg, 1965). The lipid component of this membrane is composed primarily of phospholipids (Ames, 1968), the physical properties of which are greatly determined by the structure of the constituent fatty acids (van Deenen et al, 1962; van Deenen, 1966; Chapman, 1966; Chapman et al, 1966, 1967). It was the initial purpose of the present work to isolate fatty acid auxotrophs of *Escherichia coli*, grow them on different fatty acids, and then investigate the effect of membrane lipid composition on physiology of the cell. In the course of these studies we investigated the effect of exogenous unsaturated fatty acid structure on the fatty acid composition of the membrane lipids of the auxotrophs. Our results suggest that a control mechanism is operative in *E. coli* which regulates the relative content of unsaturated and saturated fatty acids appearing in the phospholipids. These findings further suggest that as a result of this control mechanism, variations in physical properties of the lipids would be kept to a minimum even when unsaturated fatty acids of widely different structure are introduced into the lipids of the auxotrophs.

FATTY ACIDS OF *E. COLI*

Extraction of whole cells of *E. coli* with organic solvents removes all the lipids except those that are constituents of cell wall lipopolysaccharides (Kaneshiro and Marr, 1961). Lipids isolated this way are called "extractable" lipids and constitute the cell membrane phospholipids, neutral lipids, and free fatty acids (Okuyama, 1969). Fatty acid composition of "extractable" lipids shows about 45% saturated fatty acids and the rest unsaturated acids and their cyclopropane derivatives (Silbert et al, 1968; Kaneshiro and Marr, 1961). The major saturated fatty acid of this fraction is palmitic acid (40%) with minor amounts of myristic and stearic acids (Silbert et al, 1968; Kaneshiro and Marr, 1961). Unsaturated fatty acids are palmitoleic and *cis*-vaccenic acids (Norris et al, 1964; Pugh et al, 1966). Almost 85% of the fatty acids in "extractable" lipids are found in phospholipids, 5% remain as free fatty acids and the rest are in glycerides (Okuyama, 1969).

Exponentially growing cells contain none or traces of cyclopropane acids. As the culture approaches the end of logarithmic phase of growth, there is a gradual conversion of unsaturated fatty acids to their cyclopropane derivatives so that in the stationary phase of growth a large proportion of cyclopropane acids are found in the cell (Cronan, 1968).

If cells of *E. coli* are first subjected to acid or alkaline hydrolysis and then extracted with organic solvents, a substantial amount of medium chain saturated acids, $C_{12:0}$, and $C_{14:0}$, and 3-hydroxy acids are found (Marr and Ingraham, 1962; Burton and Carter, 1964). These acids are constituents of cell wall lipopolysaccharides (Burton and Carter, 1964).

The relative proportions of saturated and unsaturated fatty acids in E. coli change with growth temperature. Marr and Ingraham (1962) grew E. coli in a temperature range between 10° to 43° and noticed a gradual increase in the percentage of palmitic acid as growth temperature was increased from 10° to 43°; at the same time there was a concurrent decrease in the level of unsaturated acids with rise in temperature. Likewise, when cultures of E. coli, growing exponentially at 37°, were incubated at 10° (Okuyama, 1969), a drop in the percentage of saturated fatty acids in phospholipids with an increase in the percentage of cis-vaccenic acid was detected. The total amount of palmitic acid in phospholipids did not change at 10°; rather, there was a two-fold increase in the absolute amount of cis-vaccenic. These observations suggest that the organism maintains proper fluidity of the membrane at low temperature by incorporating higher proportions of unsaturated acids into lipids.

RESPONSE OF FATTY ACID AUXOTROPHS TO FATTY ACID SUPPLEMENTS

The growth requirements of unsaturated fatty acid auxotrophs of E. coli can be met by a variety of unsaturated fatty acids (Silbert et al, 1968; Esfahani et al, 1969; Shairer and Overath, 1969). The rate of growth, however, is not the same for all the fatty acid supplements (Table 1). The parent strain CR34, a derivative of E. coli K12, has a generation time of 95 min when grown on amino acids. The mutant strain OL_2 shows a somewhat longer generation time (105 min) when grown on the same medium supplemented with palmitoleate, oleate, cis-vaccenate, linoleate, or eicosadienoate. The rate of growth is diminished when elaidate, trans-vaccenate, linolenate, or eicosenoate is supplied (Table 1).

The fatty acid composition of the phospholipids of the auxotroph grown on oleate is very much similar to that of the parent strain (Table 2). It is worth noting that there is no difference in percentage of saturated versus unsaturated acids in the phospholipids of the parent strain grown in the presence or absence of oleate. This is at variance with observations on Lactobacillus plantarum, in which case a variety of exogenous unsaturated fatty acids effect a repression of the synthesis of cellular fatty acids (Henderson and McNeil, 1966; Weeks and Wakil, 1970).

TABLE 1

GROWTH RATES OF OL_2 STRAIN IN THE PRESENCE OF UNSATURATED FATTY ACID SUPPLEMENTS

Fatty acid supplement	Generation time
	min
cis-Δ^9-$C_{16:1}$	105
cis-Δ^9-$C_{18:1}$	107
trans-Δ^9-$C_{18:1}$	167
cis-Δ^{11}-$C_{18:1}$	105
trans-Δ^{11}-$C_{18:1}$	176
All *cis*-$\Delta^{9,12}$-$C_{18:2}$	105
All *cis*-$\Delta^{9,12,15}$-$C_{18:3}$	170
cis-Δ^{11}-$C_{20:1}$	176
All *cis*-$\Delta^{11,14}$-$C_{20:2}$	110

Cells were grown at 37° in 30 ml of medium in 300-ml Nephalo culture flasks. Optical densities were followed in a Klett colorimeter (blue filter). In all cases the carbon source was 0.2% amino acid mixture. Fatty acid supplements were included at a concentration of 100 mg/liter of medium (Esfahani et al, 1969).

TABLE 2

FATTY ACID COMPOSITION (PERCENTAGE) OF PHOSPHOLIPIDS OF E. COLI STRAINS CR_{34} AND $OL_{\bar{2}}$

Fatty acids in phospholipids	CR_{34} strain grown on		$OL_{\bar{2}}$ strain grown on amino acid plus oleate
	amino acids	amino acids plus oleate	
	%	%	%
$C_{14:0}$	1.2	0.5	0.6
$C_{16:0}$	43	42	42
$C_{16:1}$	11	10	11
$C_{17:0}$*	9.3	4.0	---
$C_{18:0}$	1.8	2.6	1.8
$C_{18:1}$	29	34	42
$C_{19:0}$*	3.1	4.6	---
Saturated	46	45	45
Unsaturated**	52	53	53
Unknown	2.1	1.3	2.0

The CR34 cells were harvested at the stationary phase of growth whereas the $OL_{\bar{2}}$ cells were harvested at midlogarithmic phase of growth.

*Cyclopropane acids

**Includes cyclopropane acids

When each member of a structurally homologous series of *cis*-unsaturated fatty acids serves as a growth factor for the auxotroph, the percentage of unsaturated fatty acid(s) present in phospholipids increases with increasing chain length or decreasing number of double bonds in the apolar chain of the supplement. Increasing the chain length of the supplement from *cis*-Δ^9-$C_{14:1}$ to *cis*-Δ^9-$C_{18:1}$

results in an increase in the amount of unsaturated fatty acid present in the phospholipids from 29 to 53% (Table 3).

TABLE 3

EFFECT OF EXOGENOUS UNSATURATED FATTY ACIDS ON THE PERCENTAGE OF SATURATED AND UNSATURATED FATTY ACIDS OF PHOSPHOLIPIDS OF OL_2^- STRAIN

Fatty acid added to growth medium	Fatty acids in phospholipids*		
	Saturated	Unsaturated	Unknown
	%	%	%
cis-Δ^9-$C_{14:1}$	70	29	1.4
cis-Δ^9-$C_{16:1}$	52	46	1.5
cis-Δ^9-$C_{18:1}$	45	53	2.0
All _cis_-$\Delta^{9,12}$-$C_{18:2}$	60	37	2.5
All _cis_-$\Delta^{9,12,15}$-$C_{18:3}$	66	30	4.4
cis-Δ^{11}-$C_{18:1}$	45	53	2.3
trans-Δ^{11}-$C_{18:1}$	29	69	1.6
trans-Δ^9-$C_{18:1}$	27	65	7.4

*C_{18} unsaturated fatty acids were partly degraded to the corresponding C_{16} acids by the auxotroph, but there was no chain elongation, or a change in the degree of unsaturation or _cis_, _trans_-isomerization (Esfahani et al, 1969).

Likewise, the relative amount of unsaturated fatty acids in the lipids decreases from 53% to 30% as the number of double bonds in the supplemental fatty acid increases from 1 to 3 (the same Table). Studies with synthetic phospholipids have revealed that shortening of chain length or increasing the degree of unsaturation in fatty acyl residue(s) results in increasing the fluidity of the lipid films (van Deenen et al, 1962; van Deenen, 1966) and lowering of the temperature at which gel to liquid crystalline phase changes are observed (Chapman, 1966; Chapman et al, 1967). The level at which the exogenous unsaturated fatty acids appear in phospholipids of the organism decreases in an orderly fashion in comparison to the level of the saturated fatty acids. Thus, the expanding effect of the unsaturated fatty acids is counterbalanced by the condensing effect of the increased level of saturated acids.

At 37°, *trans*-octadecenoic acids support the growth of the auxotroph and are incorporated at levels higher than the corresponding *cis* isomers (cf. Table 3). Physical properties of phospholipids containing *trans* acids are intermediate between those of fully saturated lipids and the ones with the corresponding *cis* acids (Chapman et al, 1966). Therefore, it appears that the organism compensated for the condensing effect of the *trans* fatty acids by effecting a reduced level of saturated acids in the mixed lipids.

Growth of the mutant at decreasing temperatures between 42° and 27° results in increasing amounts of unsaturated fatty acids in the phospholipids (Fig. 1). This is the same qualitative response to temperature shifts observed in wild type *E. coli* (Marr and Ingraham, 1962). The condensing effect of lower temperature, which favors closer packing of fatty acid chains (Lennarz, in press), appears to be overcome by elevated levels of unsaturated fatty acids in the membrane lipids. Since the amount of unsaturated acids which appear in the phospholipids of the auxotroph can not be regulated *via* unsaturated fatty acid biosynthesis, the temperature control over the unsaturated fatty acid contents of the phospholipids of wild type *E. coli* must, at least partially, lie in the biosynthesis of phospholipids.

cis-Eicosenoic acid can meet the unsaturated fatty acid requirement of the auxotroph. However, this acid was extensively degraded by $OL_{\bar{2}}$ cells to the corresponding C_{16} and C_{18} unsaturated acids prior to incorporation into phospholipids (cf. Table 4). Since it appeared that growth of this mutant on eicosenoic acid depended on the induction and activity of the β-oxidative pathway for fatty acid degradation (Weeks et al, 1969; Overath et al, 1969), a second mutant was isolated from strain $OL_{\bar{2}}$, and designated *civ-2-fao-6* which has lost the capacity to carry out fatty acid degradation. No degradative products of exogenous unsaturated fatty acids were detected in lipids of these cells (Esfahani et al, in press). However, when *cis*-Δ^{11}-eicosenoic acid was supplied to the mutant, only 30% of the total fatty acids in the phospholipids were unsaturated. Instead, a high level of myristic acid was found in the phospholipids (Table 4).

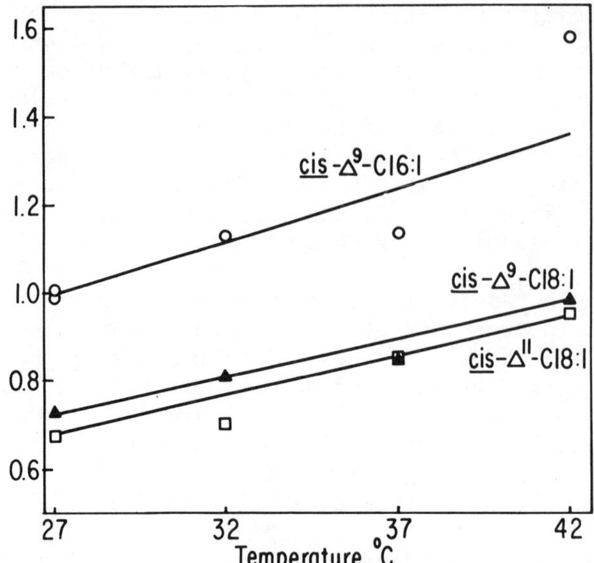

Fig. 1. Effect of temperature on the ratio of saturated to unsaturated fatty acids in phospholipids of OL_2^- strain. Cells were grown at temperatures indicated on 0.2% amino acid mixture supplemented with unsaturated fatty acids as shown. Cyclopropane acids, when present, were included with unsaturated acids for calculation.

Thus, the organism compensated for the increased chain length of the unsaturated fatty acid by introducing high levels of myristic acid into the mixed lipids. These observations indicated that the cell is able to regulate the amount and type of the saturated fatty acids synthesized in response to various growth conditions.

RELATIONSHIP BETWEEN FATTY ACID SYNTHETASE AND MEMBRANE LIPID SYNTHESIS

There is an inverse relation between the activity of the fatty acid synthetase *in vivo* and the amount of unsaturated fatty acid in the phospholipids (Esfahani et al, in press). As shown in Table 5, the amount of fatty acids synthesized from ^{14}C-acetate by resting cells was highest in cells which contain the lowest level of unsaturated fatty acid in the phospholipids.

TABLE 4

FATTY ACID COMPOSITION (PERCENTAGE) OF PHOSPHOLIPIDS OF OL_2^- AND *CIV*-2 *FAO*-6 STRAINS GROWN ON *CIS*-Δ^{11}-EICOSENOIC ACID

Fatty acids in phospholipids	OL_2^- strain	*civ*-2 *fao*-6 strain
	%	%
$C_{14:0}$	2	25
$C_{16:0}$	41	46
$C_{16:1}$	10	--
$C_{18:0}$	2	--
$C_{18:1}$	31	--
$C_{20:1}$	12	30
Unknown	2	--

A qualitative response of the synthetase to requirements for phospholipid synthesis was exhibited by experiments in which the composition of the fatty acids synthesized from ^{14}C-acetate by resting cells was determined. Cells grown on oleate have a minimal need for myristic acid. The *in vivo* products of the synthetase of such cells consisted mainly of palmitic acid (cf. Table 6) whereas, in the presence of *cis*-Δ^{11}-eicosenoic acid cells have a greater need for myristic acid, and the proportions of shorter chain acids $C_{12:0}$ and $C_{14:0}$ synthesized from ^{14}C-acetate were increased (Table 6).

The activity of the fatty acid synthetase *in vivo* thus appears to be in tune with the overall requirements of the cells for saturated fatty acids needed for phospholipid biosynthesis. Recent observations by Mindich (1970) have shown that glycerol-deprival in glycerol-requiring mutants of *Bacillus subtilis* resulted in an immediate cessation of phospholipid synthesis, and an immediate drop in the rate of fatty acid synthesis to 25% of the glycerol-supplemented control, while other cellular processes continued at normal rates for a long period of time. These independent observations confirm the conclusion of the work presented here that some sort of regulatory relationship exists between fatty acid synthetase and phospholipid synthesis.

TABLE 5

INCORPORATION OF ^{14}C-ACETATE INTO FATTY ACIDS BY RESTING CELLS OF CIV-2 FAO-6 MUTANT

Fatty acid added to growth medium	^{14}C-Fatty acid synthesized	Unsaturated fatty acids in phospholipids
	cpm/mg cells (dry weight)	% total
Elaidic	10,500	87
Oleic	17,800	59
Linolenic	42,000	37

TABLE 6

COMPOSITION OF FATTY ACIDS SYNTHESIZED FROM ^{14}C-ACETATE BY RESTING CELLS OF CIV-2 FAO-6 MUTANT

Fatty acid supplement	Fatty acid synthesized			
	$C_{12:0}$	$C_{14:0}$	$C_{16:0}$	$C_{18:0}$
		% total		
Oleic acid	7.9	11	79	2.6
cis-Δ^{11}-Eicosenoic acid	20	24	53	3.0

EFFECT OF FATTY ACID STRUCTURE AND TEMPERATURE ON SOME PHYSIOLOGICAL PROPERTIES OF THE CELL

Introduction of *trans*-octadecenoic acids into the lipids of the auxotrophs impairs several physiological functions of the cells at low, but not at high temperatures. Although these acids support logarithmic growth at 37° as shown by the exponential increase in cell number and increase in optical density, a temperature shift to 27° results in rapid loss of viability and eventual lysis (Fig. 2). The loss of viability at 27° appears to be exponential and about 90% of the original titer of viable cells are killed over a period of four hours.

The biosynthesis of both DNA and RNA were found to be adversely affected at low temperature when cells growing on elaidic acid at 37° were incubated at 30° in the presence of radioactive thymine or uracil. The synthesis of nucleic acids continued for only a short

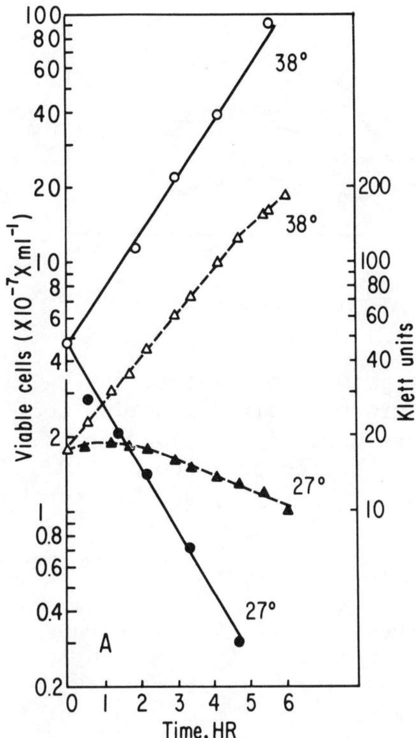

Fig. 2. Effect of temperature on growth of OL_2^- supported by elaidic acid. Cells were grown at 37° for six generations on elaidic acid with 0.2% amino acid mixture and 0.2% glycerol as carbon source. At zero time, cells, at titer shown, were incubated in the same medium at the indicated temperature. Viability was determined by serial dilution and plating on agar plates. Δ, ▲ Klett units (blue filter); O, ●, viability.

period of time at 30° (Fig. 3) while at 40° incorporation of the precursors into DNA and RNA proceeded normally, being an exponential function of growth (Fig. 3).

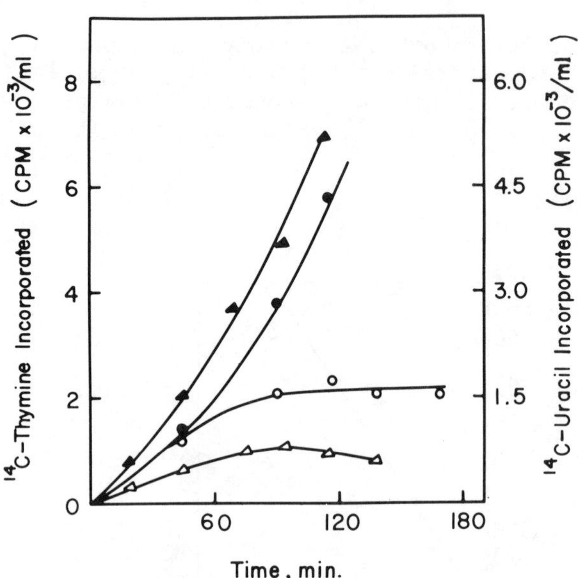

Fig. 3. Effect of growth temperature on the biosynthesis of nucleic acids by cells grown on elaidic acid. Cells growing exponentially at 37° were divided into four fractions and incubated at 40° (●,▲) or 30° (O,Δ) in the presence of 0.03 μC/ml ^{14}C-uracil, or 0.8 μC/ml ^{14}C-thymine. RNA (▲,Δ), and DNA synthesis (●,o).

Presence of trans-octadecenoic acids in phospholipids of the auxotrophs greatly affects leucine transport by chloramphenicol-treated cells at low temperature (Table 7). While no difference was detected in the initial rate of the uptake of this amino acid at 37° or 30° between cells grown on oleate or elaidate, at 18° the level of ^{14}C-leucine in elaidate-grown cells was 10% of that in cells grown on oleate. Similar observations have been made by Schairer and Overath (1969) for the transport of β-galactosides. The reduced permeability observed at low temperature in these cells is only temporary since prolonged incubations at 27° were found to cause a breakage of the permeability barrier (Esfahani et al, in press).

TABLE 7

EFFECT OF TEMPERATURE ON UPTAKE OF ^{14}C-LEUCINE BY CELLS GROWN ON OLEATE OR ELAIDATE

Fatty acid added to growth medium	^{14}C-Leucine uptake by cells at different temperatures		
	18°	30°	37°
	nmoles/g (wet weight) of cells		
Oleate	44	83	125
Elaidate	5	81	140

CONCLUSIONS AND COMMENTS

From these studies it is concluded that incorporation of fatty acids into phospholipids of *E. coli* is tightly regulated. The operation of such a regulatory mechanism enables the organism to maintain a given fluidity in the membrane lipids. Deviations from a given degree of liquidity in the lipids results in impairment of several functions in the organism as shown in cells grown on *trans*-octadecenoic acids and subjected to condensing effects of low temperature.

It has been reported that several other organisms change the degree of unsaturation in the lipids in response to various environmental conditions. As early as in 1915, Pigulewski (1915) reported that iodine value of the oil from lin seeds (*Linum usitatissium*) depended on the climate in which the plants were cultivated. The oil of seeds from Archangel had an iodine value of 190, compared to a value of 178 for those from Taganrog. Likewise, lowering of growth temperature was reported to have caused an increase in the iodine value of lipids from fungi (Terroine et al, 1935). Studies on fatty acid composition of brain lipids from goldfish acclimatized at different temperatures revealed a decrease in the percentage of saturated fatty acids, and an increase in that of polyunsaturated acids as the temperature of acclimatization was lowered (Johnston and Roots, 1964).

Yeast, grown anaerobically, is devoid of any unsaturated fatty acids. This is then compensated for by an appearance of saturated fatty acids, C_{10}, C_{12}, and C_{14} (Meyer and Bloch, 1963). Finally, van Golde and Van Deenen (1966) reported that dietary alterations induced significant changes in the fatty acid composition of rat liver membrane lipids. However, the effect of variations in membrane fatty acids was compensated for by replacement of fatty acids of related unsaturation and shifts in the relative proportions of different lecithin species.

ACKNOWLEDGMENTS

This investigation was supported in part by Grant GM-06242-12 from the National Institutes of Health, United States Public Health Service, by a grant from the Life Insurance Medical Research Fund, and by a grant from the American Cancer Society.

REFERENCES

Ames, G. 1968. Lipids of Salmonella typhimurium and Escherichia coli: structure and metabolism. J. Bacteriol. 95:833-843.

Benson, A.A. 1968. The cell membrane. A lipoprotein monolayer. p. 190-202. In L. Bolis and P.A. Pethica [Ed.], Membrane models and the formation of biological membranes. North-Holland, Amsterdam.

Burton, A.J. and H.E. Carter. 1964. Purification and characterization of the lipid A component of the lipopolysaccharides from Escherichia coli. Biochemistry 3:411-418.

Chapman, D. 1966. Liquid crystals and cell membranes. Ann. N.Y. Acad. Sci. 137:745-754.

Chapman, D., N.F. Owens and D.A. Walker. 1966. Physical studies of phospholipids. II. Monolayer studies of some synthetic 2,3-diacyl-DL-phosphatidylethanolamines and phosphatidylcholines containing trans double bonds. Biochim. Biophys. Acta 120:148-155.

Chapman, D., R.M. Williams and B.D. Ladbrooke. 1967. Physical studies of phospholipids. VI. Thermotropic and lyotropic mesomorphism of some 1,2-diacyl-phosphatidylcholines (Lecithin). Chem. Phys. Lipids 1:445-475.

Cronan, T.E., Jr. 1968. Phospholipid alterations during growth of Escherichia coli. J. Bacteriol. 95:2054-2061.

Danielli, J.F. and H. Davson. 1935. A contribution to the theory of permeability of thin films. J. Cell. Comp. Physiol. 5:495-508.

van Deenen, L.L.M. 1966. Some structural and dynamic aspects of lipids in biological membranes. Ann. N.Y. Acad. Sci. 137:717-730.

van Deenen, L.L.M., U.M.T. Houtsmuller, G.H. deHaas and E. Mulder. 1962. Monomolecular layers of synthetic phosphatides. J. Pharm. Pharmacol. 14:429-444.

Esfahani, M., E.M. Barnes, Jr. and S.J. Wakil. 1969. Control of fatty acid composition in phospholipids of *Escherichia coli*: response to fatty acid supplements in a fatty acid auxotroph. Proc. Nat. Acad. Sci. (US) 64:1057-1064.

Esfahani, M., T. Ioneda and S.J. Wakil. 1970. Studies on the control of fatty acid metabolism. III. Incorporation of fatty acids into phospholipids and regulation of fatty acid synthetase of *E. coli*. J. Biol. Chem. in press.

Fox, C.F. and E.P. Kennedy. 1965. Specific labeling and partial purification of the M protein, a component of the β-galactoside transport system of *Escherichia coli*. Proc. Nat. Acad. Sci. (US) 54:891-899.

Ganeson, A.T. and J. Lederberg. 1965. A cell-membrane bound fraction of bacterial DNA. Biochem. Biophys. Res. Comm. 18:824-835.

van Golde, L.M.G. and L.L.M. van Deenen. 1966. The effect of dietary fat on the molecular species of lecithin from rat liver. Biochim. Biophys. Acta 125:496-509.

Green, D.E., D.W. Allman, E. Bachmann, H. Baum, K. Kopazyk, E.F. Korman, S. Lipton, D.H. MacLennan, D.G. McConnell, J.F. Perdue, J.S. Rieske and A. Tzagoloff. 1967. Formation of membranes by repeating units. Arch. Biochem. Biophys. 119:312-335.

Henderson, T.O. and J.J. McNeil. 1966. The control of fatty acid synthesis in *Lactobacillus plantarum*. Biochem. Biophys. Res. Comm. 25:662-669.

Johnston, P.V. and B.I. Roots. 1964. Brain lipid fatty acids and temperature acclimation. Comp. Biochem. Physiol. 11:303-309.

Jacob, F., S. Brenner and F. Cuzin. 1963. On the regulation of DNA replication in bacteria. Cold Spring Harbor Symp. Quant. Biol. 28:329-347.

Kaneshiro, T. and A.G. Marr. 1961. *Cis*-9,10-Methylene hexadecanoic acid from the phospholipids of *Escherichia coli*. J. Biol. Chem. 236:2615-2619.

Kundig, W. and S. Roseman. 1969. Further studies on bacterial permeases. Fed. Proceedings 28:463.

Lennarz, W.J. 1970. Bacterial lipids, p. 155-184. In S.J. Wakil [Ed], Lipid metabolism, Academic Press, New York.

Marr, A.G. 1960. Localization of enzymes in bacteria. p. 433-468. In I.C. Gunsalus and R.Y. Stainier [Ed], The bacteria, Volume I, Academic Press, New York.

Marr, A.G. and J.L. Ingraham. 1962. Effect of temperature on the composition of fatty acids in *Escherichia coli*. J. Bacteriol. 84:1260-1267.

McElhaney, R.N. and M.E. Tourtellotte. 1969. Mycoplasma membrane lipids: variations in fatty acid composition. Science 164: 433-434.

Meyer, F. and K. Bloch. 1963. Metabolism of stearolic acid in yeast. J. Biol. Chem. 238:2654-2659.

Milner, L.S. and H.R. Kaback. 1970. The role of phosphatidylglycerol in the vectorial phosphorylation of sugar by isolated bacterial membrane preparations. Proc. Nat. Acad. Sci. (US) 65: 683-690.

Mindich, L. 1970. Membrane synthesis in *Bacillus subtilis*. I. Isolation and properties of strains bearing mutations in glycerol metabolism. J. Mol. Biol. 49:415-432.

Norris, A.T., S. Matsumara and K. Block. 1964. Fatty acid synthetase and β-hydroxydecanoyl coenzyme A dehydrase from *Escherichia coli*. J. Biol. Chem. 239:3653-3662.

Okuyama, H. 1969. Phospholipid metabolism in *Escherichia coli* after a shift in temperature. Biochim. Biophys. Acta 176:125-134.

Overath, P., G. Pauli and H.U. Schairer. 1969. Fatty acid degradation in *Escherichia coli*. An inducible acyl-Co A synthetase, the mapping of *old*-mutations, and the isolation of regulatory mutants. European J. Biochem. 7:559-574.

Pigulewski, G.B. 1915. Difference in the composition of oils from species of the same family (in Russian). Zh. Russ. Fiz-Khim. Obshehest. 47:393-405.

Pugh, E.L., F. Sauer, M.B. Waite, R.E. Toomey and S.J. Wakil. 1966. Studies on the mechanism of fatty acid synthesis. XIII. The role of β-hydroxy acids in the synthesis of palmitate and *cis*-vaccenate by the *Escherichia coli* enzyme system. J. Biol. Chem. 241, 2635-2643.

Robertson, J.D. 1966. Design principles of the unit membrane. p. 357-408. In G.E.W. Wolstenholme and M. O'Conner [Ed], Principles of biomolecular organization, J. and A. Churchill, London.

Salton, M.R.J. 1967. Structure and function of bacterial cell membranes. Ann. Rev. Microbiol. 21:417-442.

Schairer, H.U. and P. Overath. 1969. Lipids containing trans-unsaturated fatty acids change the temperature characteristics of thiomethylgalactoside accumulation in Escherichia coli. J. Mol. Biol. 44:209-214.

Silbert, D.F., F. Ruch and P.R. Vagelos. 1968. Fatty acid replacements in a fatty acid auxotroph of Escherichia coli. J. Bacteriol. 95:1658-1665.

Steim, J.M., M.E. Tourtellotte, J.C. Reinert, R.N. McElhaney and R.L. Rader. 1969. Calorimetric evidence for the liquid-crystalline state of lipids in a biomembrane. Proc. Nat. Acad. Sci. (US) 63:104-109.

Terroine, E.F., C. Hatterer and P. Roehring. 1935. Les acides gras des phosphatides chez les animaux poikilothermes, les vegetaux superieurs et les microorganismes. Bull. Soc. Chim. Biol. (Paris) 12:682-702.

Weeks, G., M. Shapiro, R.O. Burns and S.J. Wakil. 1969. Control of fatty acid metabolism. I. Induction of the enzymes of fatty acid oxidation in Escherichia coli. J. Bacteriol. 97:827-836.

Weeks, G. and S.J. Wakil. 1970. Studies on the control of fatty acid metabolism. II. The inhibition of fatty acid synthesis in Lactobacillus plantarum by exogenous fatty acid. J. Biol. Chem. 245:1913-1921.

MICROBIAL ADAPTATION TO EXTREMES OF TEMPERATURE AND pH

Thomas D. Brock

Department of Microbiology, Indiana University

Bloomington, Indiana 47401

The question I would like to raise in this brief article is whether there are fundamental physical and chemical limitations on the evolution of organisms able to grow under extreme environmental conditions such as we find in polluted situations. The specific environmental factors I would like to consider are high temperature and low pH, and high temperature and low pH taken together. Temperature and hydrogen ion concentration are probably the most basic environmental factors which organisms must cope with, and both natural and polluted environments exist in which these factors exist in the extreme. The attitude and approach here is quite different than that of the physiologist or biochemist, who looks at how a given organism responds to environmental change. We are interested in how living organisms, taken as totality, respond. This means we must look at stable natural environments which have been available for colonization for millions of years, so that we know there has been time for evolution to reach an equilibrium. We thus study natural thermal and acidic springs, since these provide relatively constant environments of types which have probably been available for colonization as long as life has been present on earth. Much of our own work has been reviewed recently (Brock, 1967, 1969, 1970) so that in the present paper I will give only an outline of past studies and concentrate on recent studies which have not yet been published. To simplify the discussion, I will consider high temperature and low pH separately, and then consider the two factors together.

HIGH TEMPERATURE

The upper temperature limits for different groups of organisms are given in Table 1. As can be seen, structurally simpler organisms are able to grow at higher temperatures than structurally more complex ones. There are probably no true thermophilic multicellular plants and animals, the species found at temperatures up to 45-50°C being those recently derived from lower temperature forms which have been able to extend their range into higher temperature environments where competition is lacking. This is probably especially true of the vascular plants since those species found in thermal areas in different parts of the world are quite unrelated to each other, but are related to already established lower temperature forms. In the case of eucaryotic microorganisms, a definite upper limit of 55-60°C exists, whereas procaryotes are able to grow at considerably higher temperatures, photosynthetic procaryotes up to 73-75°C and non-photosynthetic ones (bacteria) up to 100°C.

TABLE 1

APPROXIMATE UPPER TEMPERATURE LIMITS
FOR DIFFERENT GROUPS OF ORGANISMS

Group	Approximate upper limit
Vascular plants	45°C
Mosses	50°C
Vertebrates	38°C
Invertebrates	48-50°C
Protozoa	50°C
Fungi	60°C
Eucaryotic algae	56°C
Procaryotic (blue-green) algae	73-75°C
Bacteria	>99°C

Data from Brock (1967) and unpublished

There are several important points that should be made about the microorganisms living at the highest temperatures: (1) They can be called true thermophiles, since they are usually optimally adapted to the temperatures at which they are living, growing more poorly at lower temperatures; (2) The species involved are usually cosmopolitan, the same forms being found in similar habitats around the world. Thus they are probably not recently derived from non-thermophilic members of the local flora, but long, well established types; (3) Both procaryotic and eucaryotic microorganisms are of great antiquity and hence there has been plenty of time for evolution to come to equilibrium so that if certain kinds of organisms are not found at higher temperatures it is probably because of some fundamental limitation on evolution.

Why then do we find that eucaryotic organisms are unable to evolve members able to grow at temperatures as high as procaryotes?

The existence of bacteria optimally adapted to temperatures over 90°C shows that there is nothing fundamental about life processes in general which are incompatible with high temperature. Thermostable macromolecules can be made and can be arranged in the characteristic configurations we associate with living cells. It seems to me that the most likely possibility is that eucaryotes are unable to construct intracellular membrane systems able to function at high temperatures. Membranes, in general, probably have to possess a certain fluidity in order to permit passage through them of molecules, and membranes capable of passing large molecules and macromolecules probably have to be especially fluid. Such fluidity is probably incompatible with thermostability. This is especially true of the nuclear membrane of a eucaryote, which must be so constructed that components as large as ribosomes and messenger RNA can pass out.

A membrane hypothesis can also be advanced to explain the absence of photosynthetic procaryotes (blue-green algae) at temperatures above 73-75°C. The photosynthetic membrane system of blue-green algae is structurally more complex than the plasma membrane, requiring that chlorophyll molecules be stacked in a highly ordered array. It is likely that such ordering would not be possible at very high temperatures. Fossil evidence for blue-green algae exists in rocks at least two billion years old, and these organisms have been unable to exceed the thermal barrier of 73-75°C over this vast period of time. This surely suggests that there is some fundamental physico-chemical limitation which they have been unable to overcome.

LOW pH

When we turn to low pH environments, we find an entirely different picture. Table 2 presents approximate lower pH limits for different groups of organisms.

TABLE 2

APPROXIMATE LOWER pH LIMITS FOR DIFFERENT GROUPS OF ORGANISMS

Group	Lower pH limit
Higher plants	2-4
Vertebrates	3.5-4
Invertebrates	<2
Protozoa	<2
Eucaryotic algae	∿0
Fungi	∿0
Bacteria	0-1
Procaryotic algae (blue-greens)	4-5

Data from Brock (1969) and unpublished

It must be emphasized that the data of Table 2 are not completely on firm ground, since less work has been done on low pH than on high temperature environments. When the relationships shown in this table are compared with those of Table 1, some interesting differences are seen. The most striking difference is that blue-green algae are less able to cope with low pH environments than are eucaryotic algae or higher plants, whereas blue-green algae were the best adapted photosynthetic organisms to high temperature. How do we explain this?

First of all, we must point out that organisms survive in low pH environments probably by excluding in some way the hydrogen ion from the cell. A vast array of crucial biochemicals are acid labile, including chlorophyll, ATP, and DNA. There is no way that evolutionary processes can lead to the production of an acid stable ATP molecule. Thus, if organisms living in low pH environments are to use the same biochemistry as other organisms, they must develop methods for excluding hydrogen ions from the cell. Only the outer layers of a cell, such as the cell wall and the outer cell membrane, must be exposed to hydrogen ions. On the other hand, there is no convenient way to exclude heat (air conditioning systems are not possible in unicellular organisms), and adaptation of the whole cell to high temperature must thus occur.

Considering photosynthetic microorganisms, a crucial requirement is that hydrogen ions must be kept from the chlorophyll molecule since when chlorophyll is placed in acid, the magnesium ion is eliminated and pheophytin, a photosynthetically inactive porphyrin, is formed. Electron microscopic studies on blue-green algae show that the chlorophyll-bearing photosynthetic membranes are often located near the periphery of the cell, and may in fact be continuous with the plasma membrane. Thus the photosynthetic membrane system may frequently be exposed to hydrogen ions. On the other hand, in eucaryotic algae the photosynthetic membranes are within the chloroplast, itself a membrane bound organelle, and hence are far removed from the cell periphery. It seems to me that this difference may be of considerable significance in attempting to explain why blue-green algae have a lower pH limit of 4-5 whereas at least certain kinds of eucaryotic algae can grow at pH values as low as 0.

We might carry this speculation one step further and consider whether adaptation to low pH might not have been one of the main factors in the evolution of the eucaryotic alga. If we grant that blue-green algae arose first and that they are unable by their very nature to evolve members capable of growth at low pH values, those habitats on earth of low pH would have constituted a niche available for colonization. Such habitats might thus have been the first places in which a recently evolved eucaryotic alga could have become established, since there it would not have met competition from blue-green alga.

HIGH TEMPERATURE AND LOW pH

We have seen that at least some kinds of organisms can adapt to temperatures as high as boiling and pH values as low as 0. When we examine hot acid environments we find that further limitations are placed on evolutionary processes. We find that the upper temperature limit for both bacteria and algae is lower in acidic than in neutral environments. For instance, the upper temperature limit for algae at neutral pH is 73-75°C (the alga is the blue-green *Synechococcus lividus*), whereas in acid pH, it is 55-56°C (the alga is the eucaryote *Cyanidium caldarium* (Doemel and Brock, 1970). With bacteria, the upper temperature limit is progressively lower as the pH is lower (Brock and Darland, 1970). Interestingly, in the hottest, most acid environments (pH 1-2, temperature 65-85°C), the bacteria present are all devoid of cell walls, and the cell envelope consists only of a simple plasma membrane. Organisms of this kind are classified as mycoplasmas, with the name *Thermoplasma acidophila* (Darland, *et al*, submitted). Why are all the bacteria living in hot acid environments devoid of cell walls? We speculate that the β-glycosidic bond is too acid-labile to remain intact under these conditions. Cell walls in general have β-glycosidic linkages which are critical structures promoting stability, and the cell wall must by its very nature be exposed to the high hydrogen ion concentration of the medium.

AFTERWARD

The thesis of this paper has been that there are inherent chemical limitations, due to the lability of certain kinds of biochemicals or biochemical linkages, which it is impossible for organisms to overcome by evolutionary processes. I have illustrated this thesis with data derived from studies on high temperature and low pH environments. Unfortunately, much of the chemical data needed to flesh out this thesis are unavailable, but the biological observations are relatively clear cut and permit us to feel assured that the chemical speculations are reasonable. There are other extreme environments which have not received sufficient study to comment on but seem likely to provide suitable material for detailed study. These include especially habitats with high amounts of heavy metals such as copper, iron, and aluminum.

I hope this brief paper might encourage chemists to turn to a study of the inherent chemical limitations which organisms must face in adapting to extreme and polluted environments, and provide us with the data we need to place on a firmer basis the speculations advanced in this paper.

ACKNOWLEDGMENTS

The work of the author reported in this paper was supported by research grants from the National Science Foundation (GB-7815 and GB-19138) and a research contract from the Atomic Energy Commission (COO-1804-21).

REFERENCES

Brock, T.D. 1967. Life at high temperatures. Science 158: 1012-1019.

Brock, T.D. 1969. Microbial growth under extreme conditions. p. 15-41. In Microbial growth, Nineteenth symposium of the Society for General Microbiology, Cambridge Univ. Press, Cambridge

Brock, T.D. 1970. High temperature systems. Annual Review of Ecology and Systematics 1: in press.

Doemel, W.N. and T.D. Brock. 1970. The upper temperature limit of *Cyanidium caldarium*. Arch. f. Mikrobiol., in press.

Brock, T.D. and G. Darland. 1970. The limits of microbial existence: temperature and pH. Science, in press.

Darland, G., T.D. Brock, W. Samsonoff, and S.F. Conti. A thermophilic acidophilic mycoplasma, isolated from a coal refuse pile. Science, submitted.

PHOTOCHEMISTRY AND PHOTOBIOLOGY OF SPORE-FORMING BACTERIA

J. E. Donnellan, Jr., and R. S. Stafford

Biology Division, Oak Ridge National Laboratory

Oak Ridge, Tennessee 37830

Ten years ago, Beukers et al. (1959) and Wacker et al. (1962) observed that the pyrimidine base, thymine, could form dimers when irradiated with ultraviolet light (UV) either in frozen solution or when the thymine had been incorporated into the DNA of living cells. Figure 1 shows such a dimer in one chain of the DNA helix. Dimerization occurs when the excited complex of two adjacent pyrimidines causes the formation of a stable cyclobutane ring. Fortunately, thymine can be labeled easily with radioactivity and the dimers may

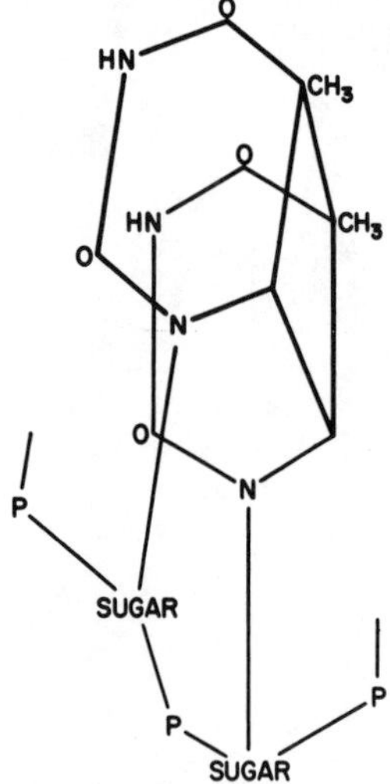

Fig. 1. Cyclobutane thymine dimer in one strand of a DNA helix. (From Setlow, 1966)

be quantiated by acid hydrolysis and chromatography. Using these and other photochemical and photobiological techniques, the Setlows (see Setlow, 1960, for review) showed that the dimers caused 50 to 90% of the lethal effect of UV in viruses and bacteria. This effect can be partially circumvented by two different types of enzymatic repair. Shortly before the discovery of cyclobutane dimers, Kelner (1951) found that certain bacteria exposed to a lethal dose of light of wavelengths around 260 nm could be reactivated by exposure to light in the region of 400 nm. Cook (1967) then showed that this photoreactivation occurred by the enzymatic monomerization of the dimerized pyrimidines back to the normal base in the presence of light. The second repair mechanism occurs in cells maintained in the dark and consists of the removal of a short segment of one strand of DNA containing the dimer and resynthesis of that segment using the complementary strand as template. Cells lacking this type of repair are very sensitive to UV as one might expect (Setlow and Carrier, 1964; Boyce and Howard-Flanders, 1964).

In the system we will describe there is evidence supporting both these repair processes, as well as processes involving a different photoproduct (Donnellan and Stafford, 1968). This system involves the spore-forming bacterium, *Bacillus megaterium*. Certain bacteria, upon reaching the stationary phase of growth, produce a dormant form that undergoes little metabolism and is very resistant to adverse environmental conditions, such as UV irradiation (Fig. 2). Given a suitable environment, the spores will germinate and produce the more sensitive vegetative cells. Vegetative cells are like most cells studied and possess the repair mechanisms for pyrimidine dimers described above. In spores, however, pyrimidine dimers are not formed and it is the response of spores to ultraviolet irradiation that we describe in this paper. First, we present a little of what we know about the photochemistry of the spore, and then describe the repair processes observed in the spore and in the vegetative cells that it produces. We also show that the product formed is involved in the lethal action of UV on spores. This leads us to consider a state in which no measurable photoproducts are formed and about which we may only speculate.

SPORE PHOTOCHEMISTRY

To study spore photochemistry we label the spores with tritiated thymidine and expose them in aqueous suspensions to ultraviolet light. The DNA is then extracted and hydrolyzed with formic acid at 175°C, or the whole spores may be hydrolyzed. The hydrolysates are chromatographed and the position of radioactivity determined.

Figure 3 shows the results of chromatograms of DNA extracted from spores and vegetative cells. At the high dose given to the

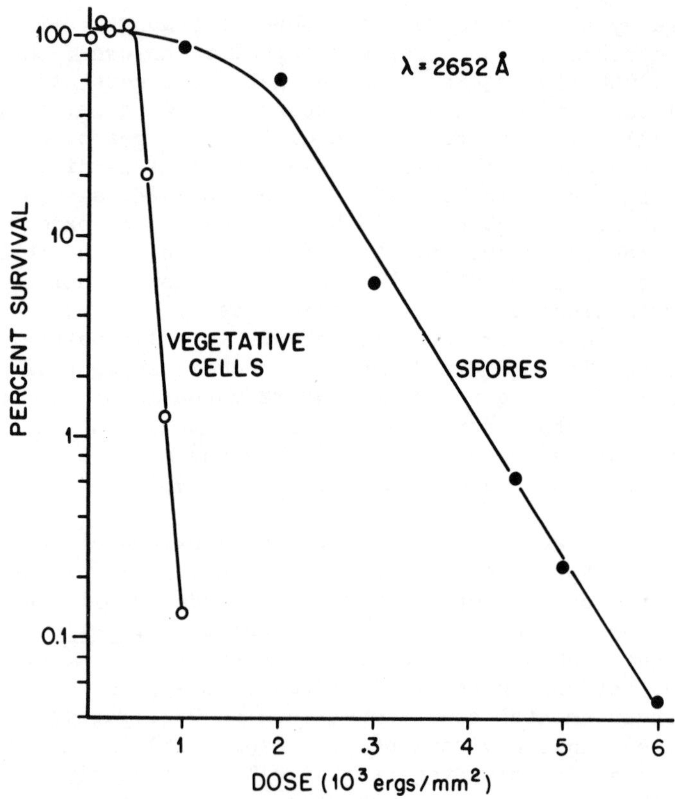

Fig. 2. Survival of B. megaterium spores or vegetative cells after 265-nm irradiation. (From Donnellan and Stafford, 1968)

spores the small amount of radioactivity in the region of the thymine-thymine dimer can be attributed to the small fraction of vegetative cells present in the population; dimers are not formed in spores. The small amounts of photoproducts other than the major one, peak b, have not been identified (note the semi-log scale). The peak labeled $\widehat{UT}$ in the distribution for vegetative cells is the radioactivity associated with the uracil-thymine cyclobutane dimer. This peak results from the deamination of the cytosine-thymine dimer during acid hydrolysis. Figure 4 depicts the kinetics of photoproduct production for both the spore photoproduct in spores and cyclobutane dimers in vegetative cells of B. megaterium. The large amount of spore photoproduct observed, about 40% of the total thymine, is close to the number of adjacent thymines in the DNA of this bacterium. We will see in a moment that this photoproduct is indeed a dimer. The fact that the cyclobutane dimers do not reach this level evolves from the photochemistry of cyclobutane dimers and is the subject of another study.

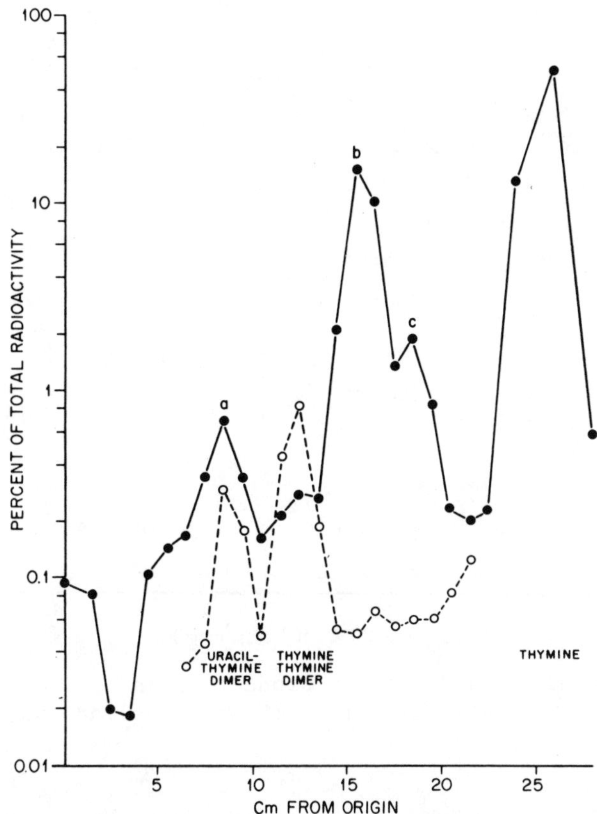

Fig. 3. Distribution of thymine radioactivity on chromatograms of hydrolyzed vegetative cells (---) and hydrolyzed DNA extracted from irradiated spores (—). (From Donnellan and Setlow, 1965)

Just as the original workers in thymine photochemistry were fortunate in being able to label the thymine dimer, we were fortunate in being able to produce the spore photoproduct in dry films of purified DNA, in frozen DNA and in dry thymidine monophosphate. Varghese (1970) has used frozen DNA and we have used dry TMP to purify the product sufficiently for characterization as shown in Figure 5. Out of habit we refer to this product as the <u>spore photoproduct</u> and use <u>dimer</u> to mean the cyclobutane dimer. The chemistry involved in these studies is outside the realm of this symposium. It is interesting to note, however, that studies using DNA labeled in the methyl group or at the 2 position of thymine demand the direct transfer of a proton (or tritium nucleus) from the methyl group of the left-hand ring to the 6 position in the right-hand ring. Presumably, this results from the fact that in dry DNA or

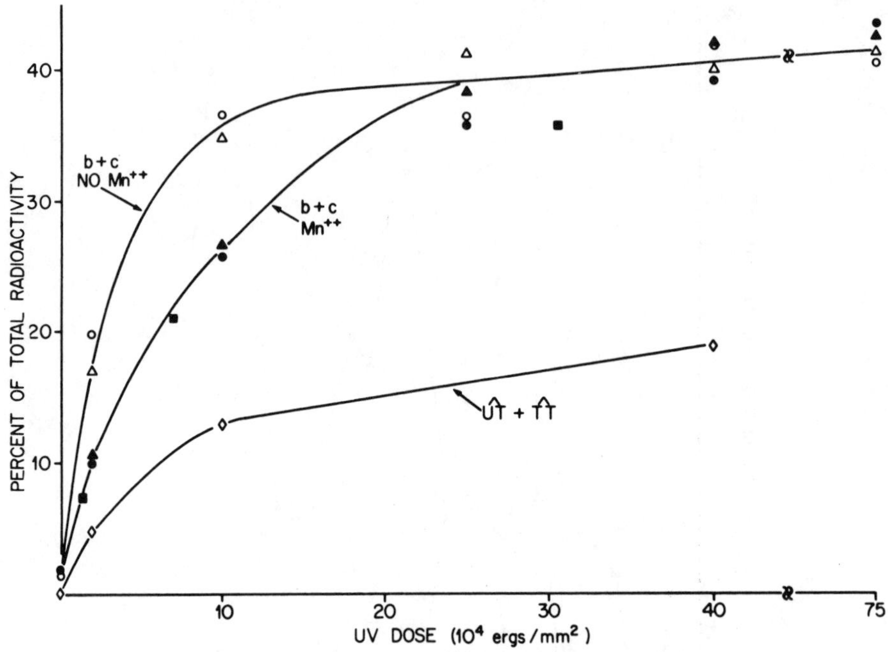

Fig. 4. Kinetics of photoproduct production in spores (b+c) or vegetative cells ($\widehat{UT}$ + $\widehat{TT}$). (From Donnellan and Stafford, 1968)

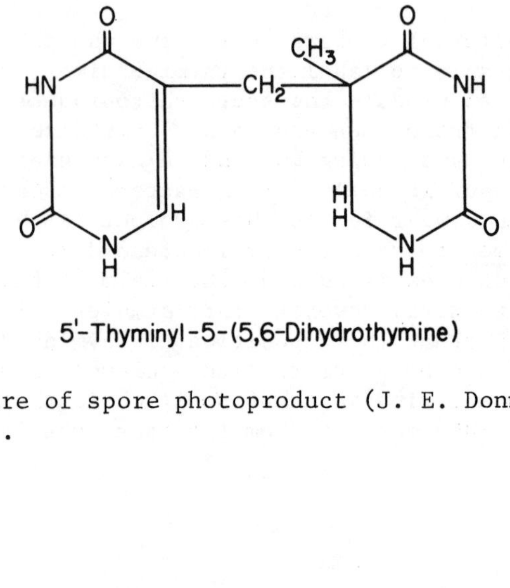

5'-Thyminyl-5-(5,6-Dihydrothymine)

Fig. 5. Structure of spore photoproduct (J. E. Donnellan, unpublished result).

dry mononucleotides, and in frozen DNA and maybe in bacterial spores in aqueous suspension, one is dealing with solid-state photochemistry.

REPAIR IN SPORES AND VEGETATIVE CELLS

Figure 6 shows an example of the photoreactivation type of repair in irradiated vegetative cells. The cells were irradiated with 265-nm light to a survival of about 5%. The sample was then irradiated at 37°C with light of 405 nm wavelength. The increase in surviving cells can clearly be observed. Table 1 shows what happens to the cyclobutane dimers present in these vegetative cells after irradiation to 5% survival. The top line, the control, shows that essentially all the radioactivity associated with dimers is in the cells' DNA, here defined as the acid-insoluble fraction. Very little of the radioactivity in the acid-soluble fraction, the small-molecule fraction, appears as dimer. The bottom line, on the other hand,

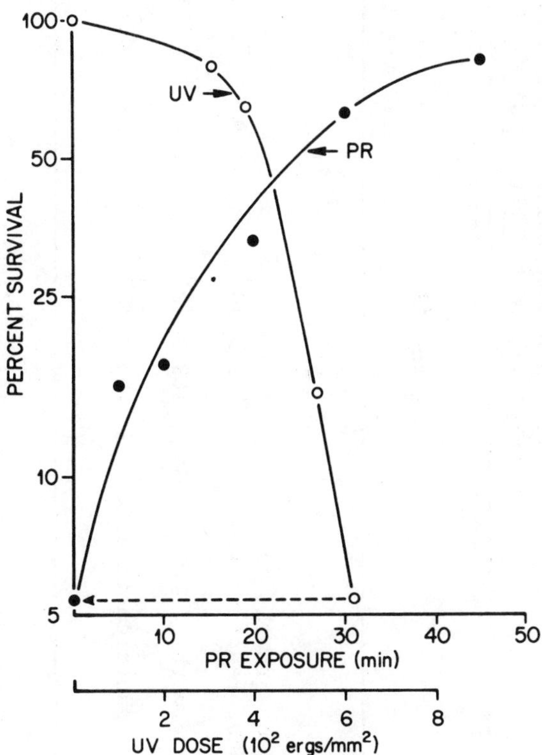

Fig. 6. Survival and photoreactivation of vegetative cells. (From Donnellan and Stafford, 1968)

TABLE 1

EXCISION AND MONOMERIZATION OF THYMINE-CONTAINING DIMERS ($\hat{UT}$ + $\hat{TT}$) IN *BACILLUS MEGATERIUM* VEGETATIVE CELLS[a]

	Total ^{3}H-T	$\hat{UT}$ and $\hat{TT}$ Acid-Insoluble		$\hat{UT}$ and $\hat{TT}$ Acid-Soluble		Total
	cpm	cpm	%	cpm	%	%
UV, no growth, no photoreactivation	139×10^3	459	0.33	52	0.04	0.37
UV, growth	169×10^3	204	0.12	375	0.22	0.34
UV + photoreactivation	153×10^3	52	0.03	70	0.05	0.08

[a]From Donnellan and Stafford (1968).

shows that photoreactivation--while increasing survival as in the previous figure--leads to a decrease in cyclobutane dimers in the cell. This result is similar to more extensive studies which show that this decrease is a result of monomerization of the dimers. We have not included any data on photoreactivation attempts on UV-irradiated spores since the spore photoproduct is not a substrate for the photoreactivation enzyme, either _in vivo_ or _in vitro_. As a consequence, exposure of irradiated spores to photoreactivating light has no effect on their survival in contrast to the results we have seen with their vegetative cells. Moreover, when irradiated spores are allowed to germinate before photoreactivation is attempted, no effect is seen. The middle line of Table 1 is an example of the dark repair mechanism we mentioned earlier. After incubation of the vegetative cells in growth medium for 25 minutes, the dimer radioactivity has moved quantitatively from the DNA fraction to the small-molecule fraction. Experiments on other strains of bacteria show that this process is accompanied by a repair synthesis of the DNA

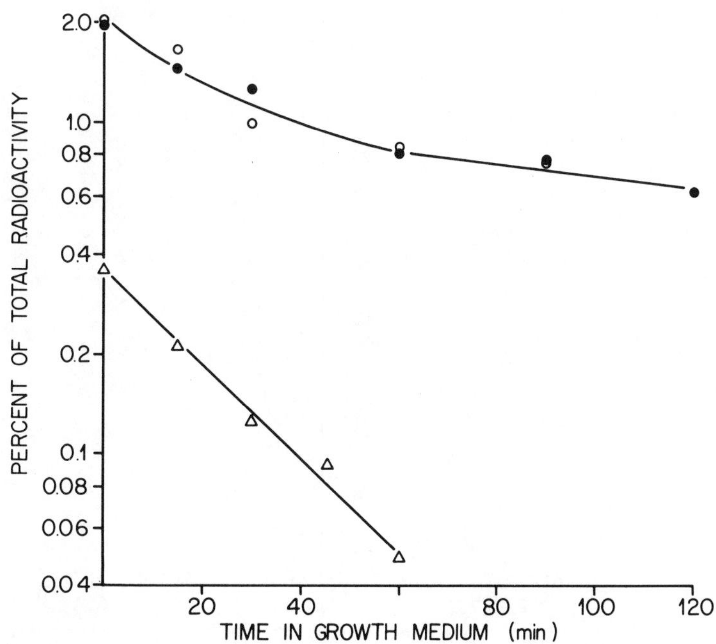

Fig. 7. Excision of dimers from vegetative cells (triangles) and disappearance of spore photoproduct from spores (circles). (From Donnellan and Stafford, 1968)

(Pettijohn and Hanawalt, 1964). The lower portion of Figure 7 shows the kinetics of this loss of radioactivity from the acid-insoluble fraction of the cells. In addition, the closed circles in the upper portion of the figure show the decrease of radioactivity in the spore photoproduct, again from the acid-insoluble fraction, when irradiated spores are allowed to germinate. In this case the spore photoproduct did not appear in the small-molecule fraction of the resulting cells nor in the medium in which they germinated. When this experiment was repeated without fractionation of the cells the results shown by the open circles resulted. In this experiment the spores were germinated in a minimal medium suitable for hydrolysis and chromatography, and the whole sample was used without separation into acid-insoluble and soluble fractions. From this we conclude that the spore photoproduct is repaired by a process different from the excision repair used by vegetative cells. Our guess is that the spore photoproduct is monomerized back to thymine.

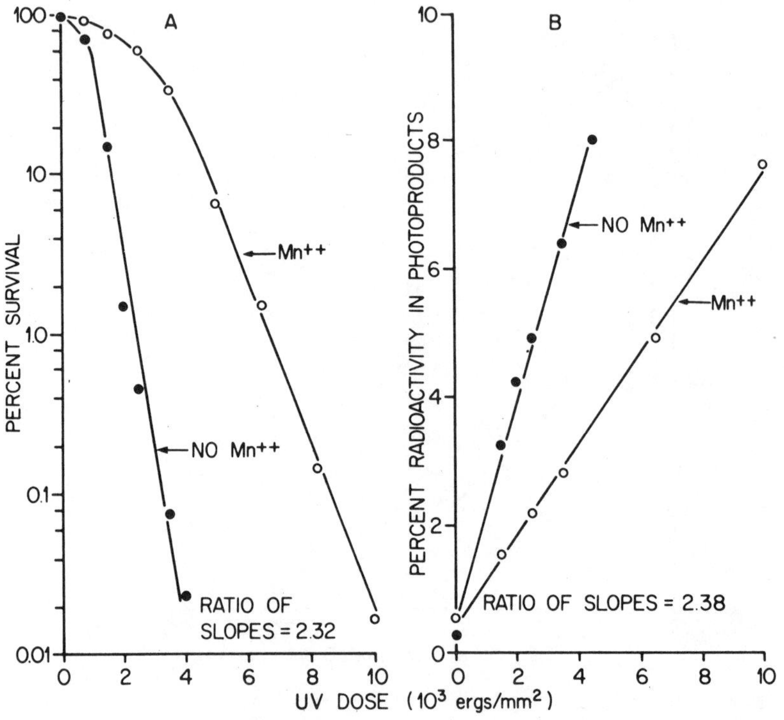

Fig. 8. Effect of added Mn^{++} on survival and photoproduct production in spores. (From Donnellan and Stafford, 1968)

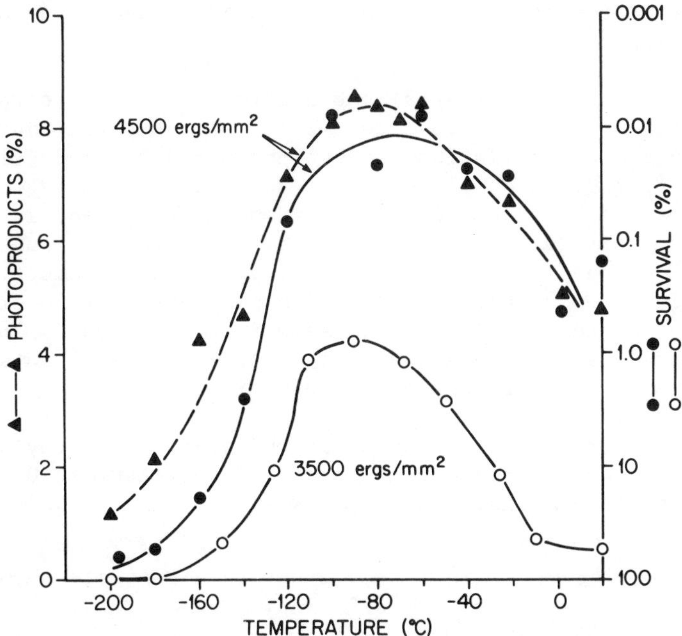

Fig. 9. Survival and photoproduct production of spores as a function of temperature during irradiation. (From Donnellan et al., 1968)

LETHALITY OF SPORE PHOTOPRODUCT

In order to interpret the last result as a repair mechanism we must be able to show that the spore products are correlated with cell death, usually a difficult task. Here again nature has been kind to us. Our usual method for producing spores is to place some vegetative cells in a liver extract medium and allow nature to take its course. Liver extract contains a wealth of different chemicals, but if it is supplemented with a few millimoles of manganous ion the radiation sensitivity of the resulting spores is dramatically altered (Fig. 8). We would like to emphasize that excess manganese was present only during the production of spores. After spore formation the culture was washed many times with water and the irradiations were performed on an aqueous suspension of spores. The samples used to determine survival were also analyzed for spore photoproduct and these results are shown on the right of the figure. In spores grown with excess manganese the spore photoproduct is less readily produced than in spores grown without manganese, and survival is subsequently higher. In fact the ratio of the slopes in both cases are in remarkably good agreement.

The next experiment (Fig. 9) shows a remarkable bit of photochemistry which we are at a loss to explain on a physical basis but which again shows a correlation between photoproduct production and lethality of spores. The solid circles show the lethality to spores of a fixed dose of UV as a function of temperature during irradiatior In this graph, survival decreases upwards. The solid triangles show a concomitant change in the production of spore photoproduct. The lower curve is a repeat of the survival experiment at a lower dose. The dependence on temperature is the same in either case. In one experiment at 4° Kelvin the extrapolation of an increase in survival and a decrease in photoproduct holds very nicely.

The last experiment is based on an observation of Stuy's (1956) which suggested that spores become even more resistant to UV before they form vegetative cells. Since germination is very rapid and

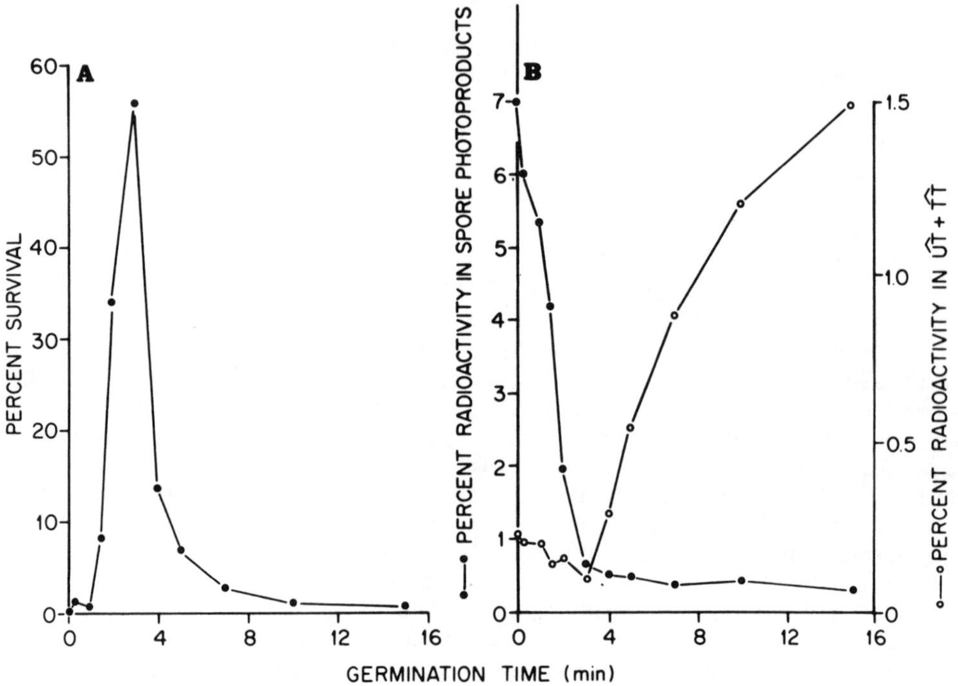

Fig. 10. Survival and photoproduct production after UV radiation given at different times during germination of _B. megaterium_ spores. (From Stafford and Donnellan, 1968)

synchronous in B. megaterium, it is easier to show this result as the survival of the population as a function of time in germination medium after a fixed dose of radiation (Fig. 10). On the left of the figure we see that a dose of UV which permitted less than 1% survival of dormant spores allowed 55% survival of the population 3 minutes after the start of germination. The survival subsequently decreases as vegetative cells are formed. The right-hand part of this graph shows the production of spore photoproduct decreasing rapidly after the start of germination while production of cyclobutane dimers is delayed. Thus germinating spores are very resistant to UV because neither spore photoproducts nor cyclobutane dimers are produced. The different scales show that cyclobutane dimers are some ten times more lethal than the spore photoproduct.

We have seen in this paper several examples of how bacteria respond to the insult of ultraviolet light. They may repair the lesions either by photoreactivation or dark repair or, in the case of spore-forming bacteria, they may adapt by altering the structure of their DNA so that fewer lethal photoproducts are produced or none at all.

ACKNOWLEDGMENTS

Our research was sponsored by the U. S. Atomic Energy Commission under contract with Union Carbide Corporation.

REFERENCES

Beukers, R., J. Ylstra and W. Berends. 1958. The effect of ultraviolet light on some components of the nucleic acids. Rec. Trav. Chim. 77: 729-732.

Boyce, R.P. and P. Howard-Flanders. 1964. Release of ultraviolet light-induced thymine dimers from DNA in E. coli K-12. Proc. Nat. Acad. Sci. (US) 51: 293-300.

Cook, J.S. 1967. Direct demonstration of the monomerization of thymine-containing dimers in U.V.-irradiated DNA by yeast photoreactivating enzyme and light. Photochem. Photobiol. 6: 97-101.

Donnellan, J.E., Jr., J.L. Hosszu, R.O. Rahn and R.S. Stafford. 1968. Effect of temperature on the photobiology and photochemistry of bacterial spores. Nature 219: 964-965.

Donnellan, J.E., Jr. and R.B. Setlow. 1965. Thymine photoproducts but not thymine dimers found in ultraviolet-irradiated bacterial spores. Science 149: 308-310.

Donnellan, J.E., Jr. and R.S. Stafford. 1968. The ultraviolet photochemistry and photobiology of vegetative cells and spores of Bacillus megaterium. Biophys. J. 8: 17-28.

Kelner, A. 1951. Action spectra for photoreactivation of ultraviolet-irradiated Escherichia coli and Streptomyces griseus. J. Gen. Physiol. 34: 835-852.

Pettijohn, D. and P. Hanawalt. 1964. Evidence for repair-replication of ultraviolet damaged DNA in bacteria. J. Mol. Biol. 9: 395-410.

Setlow, J.K. 1966. The molecular basis of biological effects of ultraviolet radiation and photoreactivation, p. 195-248. In M. Ebert and A. Howard [Ed], Current topics in radiation research, Volume II, North Holland, Amsterdam.

Setlow, R.B. 1966. Cyclobutane-type pyrimidine dimers in polynucleotides. Science 153: 379-386.

Setlow, R.B. and W.L. Carrier. 1964. The disappearance of thymine dimers from DNA: an error correcting mechanism. Proc. Nat. Acad. Sci. (US) 51: 226-231.

Stafford, R.S. and J.E. Donnellan, Jr. 1968. Photochemical evidence for confirmation changes in DNA during germination of bacterial spores. Proc. Nat. Acad. Sci. (US) 59: 822-828.

Stuy, J.H. 1956. Studies on the mechanism of radiation inactivation of micro-organisms. III. Inactivation of germinating spores of Bacillus cereus. Biochim. Biophys. Acta 22: 241-246.

Varghese, A.J. 1970. 5-thyminyl-5,6-dihydrothymine from DNA irradiated with ultraviolet light. Biochem. Biophys. Res. Comm. 38: 484-490.

Wacker, A., H. Dellweg and D. Jacherts. 1962. Thymin-dimerisierung und Überlebensrate bei Bakterien. J. Mol. Biol. 4: 410-412.

INDUCTION OF THE HEPATIC AMINO ACID TRANSPORT SYSTEM AND TYROSINE AMINOTRANSFERASE IN RATS ON CONTROLLED FEEDING SCHEDULES

David F. Scott, Fred R. Butcher, Robert D. Reynolds and Van R. Potter

McArdle Laboratory, University of Wisconsin Medical School, Madison, Wisconsin 53706

In considering the "biochemical responses to environmental stress" in mammalian organisms it is appropriate to examine the adaptive changes that occur in the liver. This organ is not only concerned with the detoxification of toxic environmental hazards but it is also primarily responsible for maintaining homeostasis in the internal environment despite uneven variations in the time of food ingestion and in the proportions of dietary carbohydrate, protein and fat. The concept of environmental stress is coupled with the concept of physiological adaptation and both have been given definitions that range from the idea that stress is something harmful to the idea that some stress and adaptation is part of everyday living and that indeed we have to inquire about the range of stress levels that might be regarded as part of an "optimum environment" (see Aschoff, 1967, The Handbook of Physiology, Sect. 4 and discussions by Potter, 1969, 1970).

The studies presented here were performed using rats adapted to what we have called the "8+16" controlled feeding schedule in which rats are allowed access to food for only the first 8 hours of the 12-hour period of darkness which is arbitrarily scheduled from 8:30 A.M. to 8:30 P.M. each day (Potter et al., 1968). Controlled feeding schedules such as the "8+16" regimen result in an improved synchronization of the rat's metabolic activities such that diurnal variations may be observed to occur with remarkable precision and apparently greater amplitude when based on the average data from groups of adapted rats killed at different times in the 24-hour cycle: the greater amplitude could result from a better synchronization of fluctuations in the members of each group.

It is the purpose of the present report to emphasize the sensitivity and the correlation of two parameters that have been frequently studied separately by others but apparently have not been simultaneously measured except in the present and previous reports from this laboratory. We wish to advocate the adoption of both the controlled feeding and lighting schedules and also the measurement of amino acid transport in connection with biochemical studies on physiological adaptations to natural and unnatural environmental situations. Our previous work has emphasized the measurement of tyrosine aminotransferase, TAT, (EC 2.6.1.5) although studies on other enzymes and glycogen have been carried out. It has been shown that the daily fluctuations in tyrosine aminotransferase (Watanabe et al., 1968) are accompanied by fluctuations in the amino acid transport system, TS, in rats on controlled feeding schedules (Baril and Potter, 1968, Potter et al., 1968)[1].

We wish to demonstrate the sensitivity of tyrosine aminotransferase in rat liver as an indicator of fluctuations in a variety of hormonal responses to stress and to show that the amino acid transport system, as studied by means of radioactive α-aminoisobutyric acid (AIB) and/or aminocyclopentane-carboxylic acid (ACPC, also called cycloleucine), is also a sensitive indicator of the mammalian response to environmental change. These two non-metabolizable amino acids have been introduced by H. N. Christensen and shown to be actively transported but not oxidized or incorporated into protein (Noall et al., 1957; Akedo and Christensen, 1962; Christensen and Jones, 1962; Riggs et al., 1963).

In the earlier reports (Baril and Potter, 1968; Potter et al., 1968; Watanabe et al., 1968) a positive correlation between

---

[1] Although our results on the transport of cycloleucine are qualitatively the same, we have been unable to confirm the distribution ratios (liver/blood) for cycloleucine reported earlier (Baril and Potter, 1968; Potter et al., 1968) which were higher than the ratios given in the present report. Attempts to find a systematic calculation or procedural difference have been unsuccessful. Decomposition of the radioactive ACPC was suspected and possibly may occur in the solid state. However, mixed samples of ^{3}H and ^{14}C ACPC gave identical distribution ratios, and a sample of repurified ACPC obtained from H. N. Christensen (whose cooperation is gratefully acknowledged) gave the same distribution ratio as the other samples used for the present work.

the activities of TS and TAT was noted for perturbations of TAT activity by either hormonal or dietary means. This relationship was maintained in certain minimal deviation hepatomas (Baril et al., 1968, 1969; Watanabe et al., 1969). We recently extended the correlation of these processes to include their co-induction by theophylline, glucagon, and dibutyryl-cyclic-AMP (Scott et al., 1970). The present report summarizes experiments designed to examine further the interrelationship of these parameters and their response to theophylline (Fig. 1).

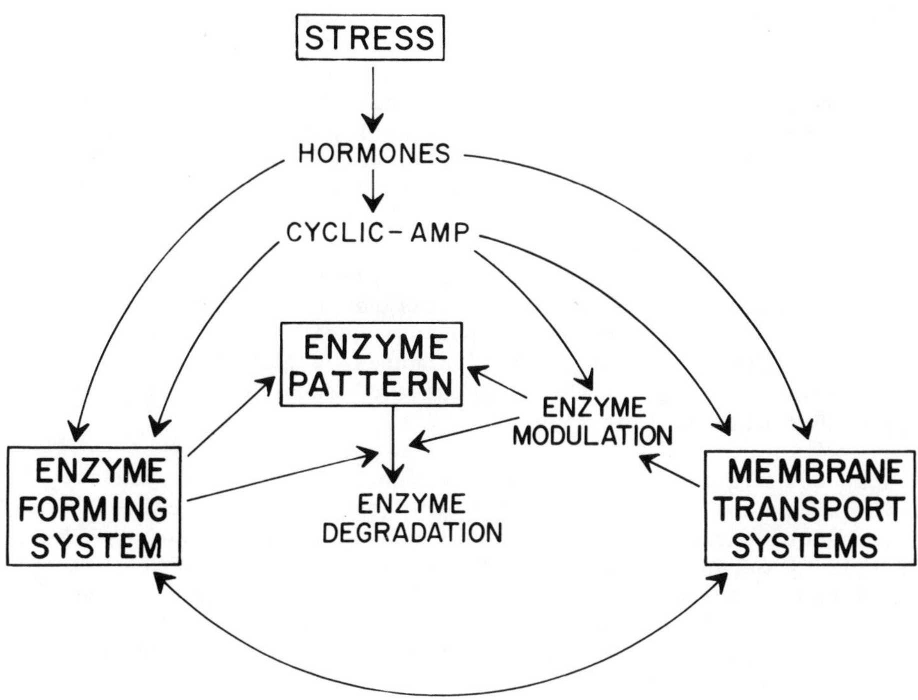

FIG. 1. Schematic representation of biochemical responses to environmental stress. Theophylline inhibits the diesterase that converts cyclic AMP to 5'-AMP although it may well have other effects (Breckenridge, 1970).

METHODS

Male rats were obtained from the Charles River Breeding Laboratories, Wilmington, Massachusetts at 22 or 36 days of age. On arrival they were housed in a windowless room with lighting regulated to provide alternating 12 hour periods of light and darkness. Food was available only during the first 8 hours of the dark period. This feeding schedule is designated as the 8+16 regimen. Rats were allowed at least one week to adapt to the feeding schedule and to chemically defined diets (Watanabe et al., 1968) in which 12, 30 and 60% casein was varied with glucose (General Biochemicals, Inc., Chagrin Falls, Ohio). Unless otherwise noted, the rats were not fed on the day of the experiment and had been without food for 16 hours after their last 8-hour feeding period.

The model amino acids which were obtained from New England Nuclear Corporation, Boston, Massachusetts, were given to the rats by subcutaneous injection (AIB-1-^{14}C, 1 μCi, 0.125 μmoles/100 g body weight and ^{3}H-ACPC, 2.5 μCi, 0.036 μmoles/100 g body weight) during the feeding period of the day preceding the experiment. The theophylline and other agents usually were dissolved in 0.9% NaCl at concentrations appropriate to give the specified dose in 1 ml/100 g body weight for intraperitoneal injection. At the indicated times after injection, the rats were killed by decapitation. After the blood was collected in heparinized beakers, the livers were removed, quickly chilled, weighed and homogenized in 4 volumes of 0.9% NaCl with a polytron homogenizer.[2] After centrifugation at 104,000 x g for 15 minutes, supernatant fractions were stored at -70°C until they were assayed for TAT activity by the method of Diamondstone (1966).

Acid soluble radioactivity was determined in the supernatant fraction after centrifugation of mixtures of 1 volume 6.1 M trichloroacetic acid and 10 volumes of plasma or homogenate. All radioactivity measurements were done using Packard Tri-Carb liquid scintillation spectrometers. The activity of TS is expressed as the ratio of the distribution of the model amino acids between the tissue and the plasma as measured by the radioactivity present in the acid soluble extracts.

---

[2] Manufactured by Kinematica GMBH, Luzen-Schwerz.

RESULTS

Effect of Feeding 60% Protein Diet and Theophylline Treatment on TS and TAT

Figure 2 shows the increases in TAT and in the transport of both amino acid analogs that are seen in rats fed the 60% diet. Although the same type of response is seen with diets of lower protein concentrations, the increases in these parameters are much less marked (Baril and Potter, 1968; Watanabe et al., 1968). TAT reaches its peak activity, several times its basal value, 4-8 hours after feeding begins. Not shown here is the gradual return to the basal level which ensues when feeding is stopped. The distribution ratios undergo similar but less marked changes, three-fold change for AIB and 0.5-fold for ACPC. The values of the distribution ratio for ACPC throughout these studies were considerably lower than those previously reported by this laboratory. (Baril and Potter, 1968; Potter et al., 1968). Riggs et al. (1963) noted ACPC ratios lower than AIB ratios and similar to those in the present report.[1]

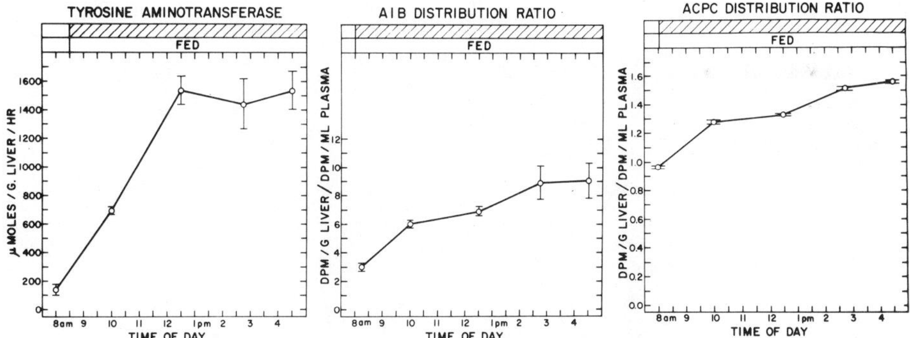

FIG. 2. Diurnal increases of TAT and TS during the feeding period in rats adapted to the "8+16" regimen with 60% protein diet. Each symbol represents the mean ± standard error from 3 or 4 rats killed at the indicated time in the cycle.

If the rats are not fed on schedule, the activities of these parameters remain at their quiescent levels. This is shown in Figure 3 which gives the time course of the responses of TAT and TS to different doses of theophylline. During the 8 hours examined, the rats receiving saline had very little perturbation of either TAT or TS. For both analogs, maximum values of TS were obtained at about 4 hours after theophylline injection, whereas maximum activity of TAT was not obtained until after 4 hours. Thus a very strong correlation between these processes was observed for the first 4 hours after which time, the transport activity was maintained at its saturation level while the enzyme still was accumulated. As expected, both the magnitude and the duration of the response of each parameter was a function of the dose. Subsequent studies showed that 12-16 hours were required before these processes returned to basal values after a dose of 10 mg/100 g body weight.

The Role of Macromolecular Synthesis in the Response of TAT and TS to Theophylline Treatment

Since a requirement of protein synthesis was well-established for adaptive changes in the activity of TAT to occur, it was of

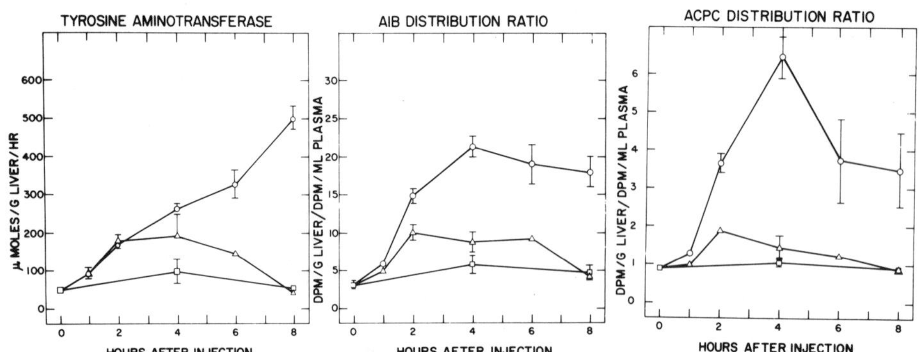

FIG. 3. Time course of co-induction of TAT and TS by theophylline in intact rats. Squares, triangles and circles respectively denote doses of 0, 5 and 10 mg/100 g body weight. The rats were injected 16 hours after their last 8-hour feeding period. Food was withheld during the experiment. Each symbol represents the mean ± standard error from 3 rats killed at the indicated times after injection.

interest to determine what effect inhibitors of protein synthesis would have on the response of TS to theophylline treatment. Figure 4 shows that puromycin at a dose of 10 mg/100 g body weight blocked most of the theophylline-induced rise in both TAT and TS. When given alone, puromycin had no discernable effect on either parameter. The small response observed in animals given both drugs may have been due to an incomplete inhibition of protein synthesis.

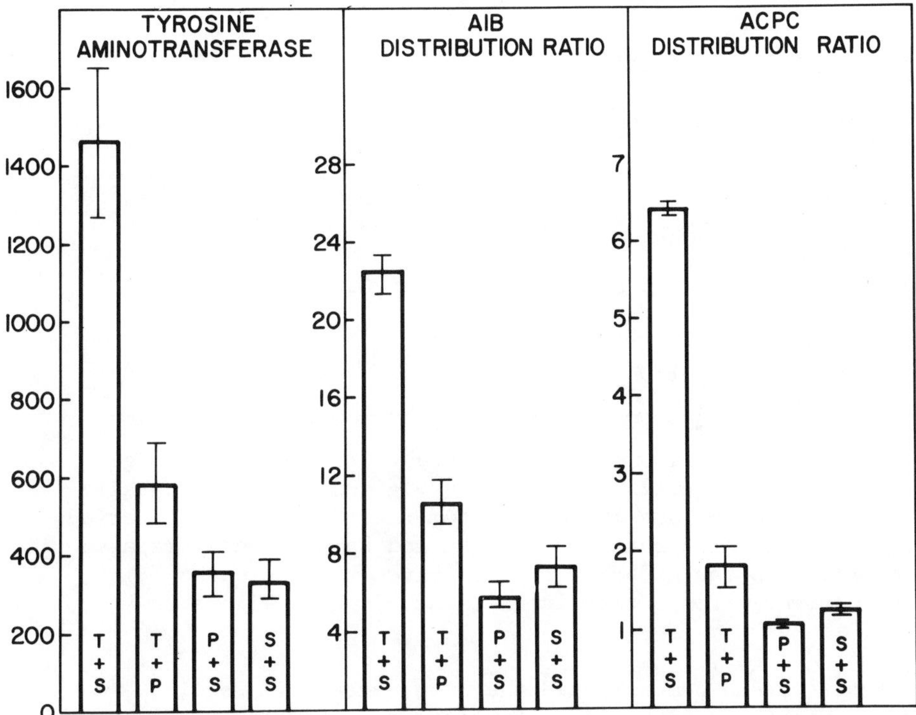

FIG. 4. Inhibition by puromycin of the induction of TAT and TS by theophylline. Saline (S, 1 ml/100 g body weight) and/or theophylline (T, 7.5 mg/100 g body weight) and puromycin (P, 10 mg/100 g body weight) were given intraperitoneally to rats 16 hours after their last feeding period. Four hours later the rats were killed and samples were prepared as described in the Methods. Each bar represents the mean ± standard error from 3 rats. Units are as defined in Fig. 2.

Additional data are presented in Table 1 which summarizes the blockade by cycloheximide of the theophylline induction of TAT and TS. In this experiment, cycloheximide completely blocked the induction of the enzyme. Values of the distribution ratios were somewhat higher in the rats receiving both drugs than in those treated with cycloheximide alone, although lower than in rats receiving saline only.

TABLE 1

EFFECTS OF CYCLOHEXIMIDE ON THE INDUCTION OF TS AND TAT BY THEOPHYLLINE *IN VIVO*

TREATMENT	TAT ACTIVITY	ACPC RATIOS	AIB RATIOS
Theophylline + Saline	688 ± 24	3.34 ± 0.61	16.59 ± 1.35
Saline + Saline	244 ± 86	1.03 ± 0.08	5.65 ± 0.88
Theophylline + Cycloheximide	165 ± 29	1.09 ± 0.02	3.83 ± 0.20
Saline + Cycloheximide	173 ± 49	0.93 ± 0.01	2.57 ± 0.19

Saline (1.5 ml/100 g) and/or theophylline (7.5 mg/100 g) and cycloheximide (0.125 mg/100 g) were given intraperitoneally to 49-day-old male rats 16 hours after their last feeding period. Four hours later the rats were killed and samples were prepared as described in the Methods. Values given are the means ± standard errors from 3 rats in units as given in Figure 2.

After we had demonstrated the involvement of protein synthesis in the theophylline stimulation of TAT and TS, it was of interest to determine what effect an inhibitor of nucleic acid synthesis, actinomycin D, might have on the response of TAT and TS to theophylline. As shown in Figure 5, treatment with actinomycin D (which inhibited by 78% the incorporation of ^{3}H-orotic acid into RNA over a 4 hour period) also blocked the theophylline-induced increases of TAT and TS. However, the inhibition of nucleic acid synthesis was less effective in blocking the response of TS to theophylline than the inhibition of protein synthesis in the previous experiment.

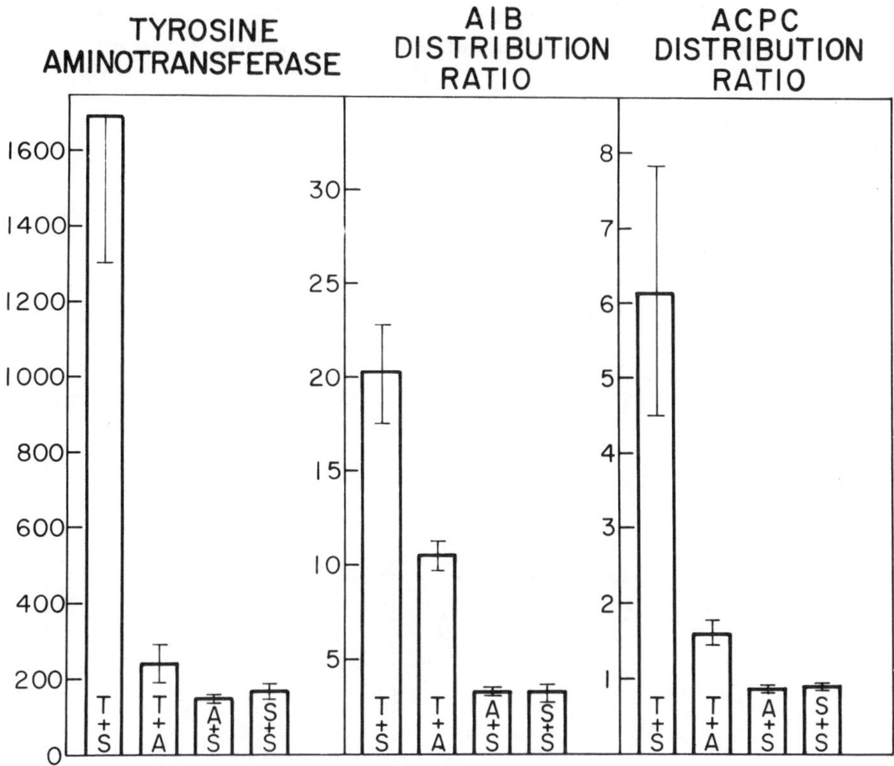

FIG. 5. Inhibition by actinomycin D of the induction of TAT and TS by theophylline. Saline (S, 1 ml/100 g body weight) and/or theophylline (T, 7.5 mg/100 g body weight) and actinomycin D (A, 0.1 mg/100 g body weight) were given intraperitoneally to rats 16 hours after their last feeding period. Four hours later the rats were killed and samples were prepared as described in the Methods. Each bar represents the mean ± standard error from 3 animals. Units are as defined in Fig. 2.

Stabilization of Induced Levels of TAT and TS after Inhibition of Protein Synthesis

In an attempt to affect differentially TAT and TS we did a "Kenney" (1967) experiment in which groups of rats whose TAT had already been induced were given either cycloheximide or saline and the decay of activity was followed. As shown in Figure 6, the inhibition of protein synthesis with cycloheximide stabilized the TAT at the induced level in agreement with Kenney's report and cycloheximide treatment resulted in a partial stabilization of TS as measured with ACPC. It appears that both processes have a requirement for protein synthesis in order to return to their basal levels.

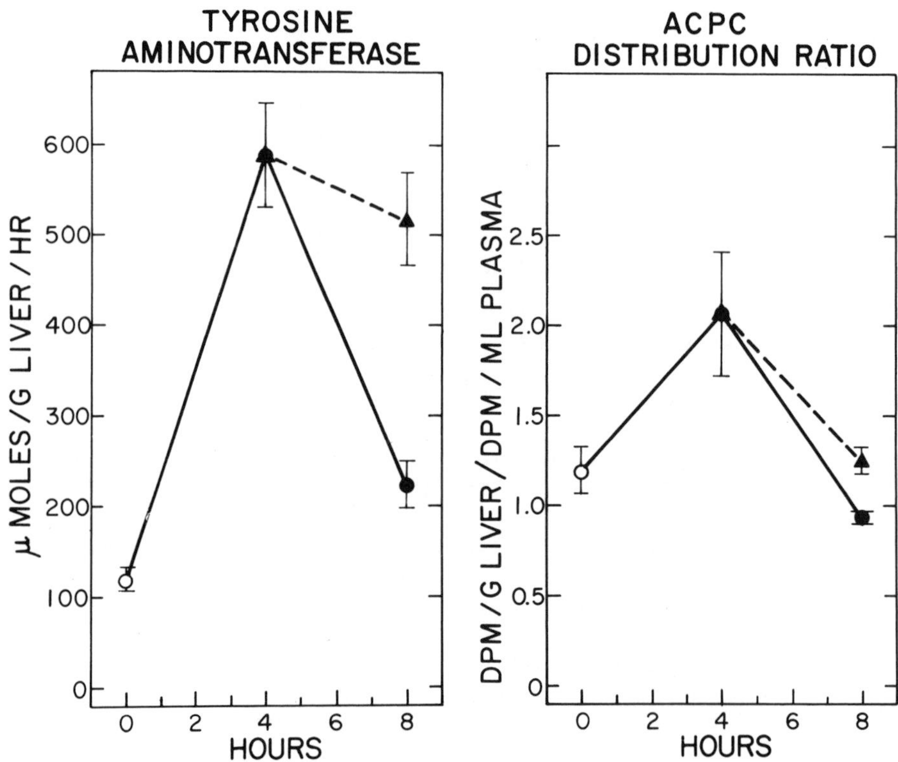

FIG. 6. Stabilization of TAT and TS after inhibition of protein synthesis. At 0-time, 16 hours after their last feeding period, the rats were given intraperitoneal injections of theophylline (7.5 mg/100g body weight). At four hours they were given either saline (1 ml/100 g body weight) denoted by the solid circles or cycloheximide (0.1 mg/100 g body weight) denoted by the solid triangles. Each symbol represents the mean ± standard error from 3 rats killed at the indicated times.

DISCUSSION

The present studies indicate that the enzyme forming system plays a key role in the stimulation of TAT and TS by theophylline. It appears that protein synthesis is a minimal requirement for the enhancement of the activities of these parameters to occur. The blockade of the effects of theophylline by actinomycin D suggests that RNA synthesis is required to obtain the necessary protein synthesis. The interruption by antimetabolites of TAT inductions has been reported for almost all of the stimuli listed in Table 2. The literature on the effects of inhibitors of protein and nucleic acid synthesis on the stimulation of the hepatic amino acid transport system is far less extensive. Chambers et al. (1965) reported that treatment with actinomycin D 4 hours prior to beginning a liver perfusion had no effect on the hydrocortisone stimulation of AIB uptake by the perfused liver, whereas actinomycin D treatment did reduce the insulin stimulation of AIB uptake to about half that observed in livers from untreated rats. More recently, Tews et al. (1970) have reported that cycloheximide and actinomycin D blocked the cyclic-AMP-induced stimulation of AIB transport in liver slices. This observation is particularly relevant since theophylline treatment results in an increase in the hepatic levels of cyclic-AMP.[3] We believe that our results on the kinetics of change of TS (Scott et al., 1970, and the present report) and on the effects of metabolic inhibitors *in vivo* in conjunction with those of Tews et al. (1970) indicate that the rate-limiting component in the hepatic amino acid transport system is a polypeptide with a short half-life.

The sensitivity of tyrosine aminotransferase to a variety of stimuli is emphasized by Table 2 in which only those papers published in 1970 are cited. In over thirty reports no less than 15 different stimuli have been examined. The hepatic amino acid transport system (see Table 3) is not nearly as popular a system for study as TAT. However, the reader is referred to the excellent reviews by Riggs (1964 and 1970) for a comprehensive treatment of the effects of hormones on the membrane transport systems of different tissues some of which have been studied more extensively than liver. Although we have recommended the hepatic amino acid transport system as a sensitive indicator of environmental stress, fluctuations in the uptake of amino acids by other tissues may be equally sensitive to stress.

---

3 Butcher, F. R., D. F. Scott and V. R. Potter, Unpublished observations from experiments in progress.

TABLE 2

INDUCTION OF TYROSINE AMINOTRANSFERASE IN RESPONSE TO ENVIRONMENTAL STRESS: SUMMARY OF 1970 REPORTS ON THE REGULATION OF HEPATIC TYROSINE AMINOTRANSFERASE *

Stimulus	Parameters studied	Reference
I. FOOD (Diurnal variation)		
A. dietary, protein	TAT	Black and Axelrod, 1970c
	TAT, TAT synthesis	Civen et al., 1970
	TAT, adrenal corticosterone	Cohn et al., 1970
	TAT, hepatic tryptophan	Fuller, 1970a
	TAT, in liver and brain	Fuller, 1970b
	TAT	Kreutler and Miller, 1970
	TAT	Mavrides and Lane, 1970
	TAT, tyrosine levels and utilization *in vivo*	Rose and Wurtman, 1970
B. amino acids	TAT	Kato, 1970a
	TAT	Mavrides and Lane, 1970
	TAT and tryptophan oxygenase	Yuwiler et al., 1970b
II. HORMONES		
A. estrogen	TAT	Rose and Cramp, 1970
B. glucagon	TAT, in liver and brain	Fuller, 1970b
	TAT, ornithine-δ-transaminase, histidine ammonia lyase, phosphoenolpyruvate carboxykinase	Nakajima et al., 1970
	TAT, serine dehydratase	Potter et al., 1970
	TAT, AIB transport	Scott et al., 1970

(TABLE 2 CONT.)

Agent	Enzyme/system	Reference
C. insulin	TAT, ornithine-δ-transaminase, histidine ammonia lyase, phosphoenolpyruvate carboxykinase	Nakajima et al., 1970
D. glucocorticoids	TAT, genetic differences	Blake, 1970a,b,c
	TAT, in liver and brain	Fuller, 1970b
	TAT	Fuller and Snoddy, 1970
	TAT, isozymes	Holt et al., 1970
	TAT, TAT synthesis	Levitan and Webb, 1970a,b
	TAT, tryptophan oxidase	Liberti et al., 1970
	TAT	Mavrides and Lane, 1970
	TAT, ornithine-δ-transaminase, histidine ammonia lyase, phosphoenolpyruvate carboxykinase	Nakajima et al., 1970
	TAT, serine dehydratase	Potter et al., 1970
	TAT, isozymes	Sadleir et al, 1970
	TAT, protein synthesis *in vitro*	Sekeris et al., 1970
	TAT	Susten and Kirksey, 1970
	TAT, tryptophan oxygenase	Yuwiler et al., 1970b
III. NEUROGENIC AGENTS		
A. carbamyl choline	TAT	Black, 1970a
B. epinephrine	TAT, in liver and brain	Fuller, 1970b
	TAT	Fuller and Snoddy, 1970
C. norepinephrine	TAT	Black and Axelrod, 1970a,b,c
	TAT	Black, 1970b
D. reserpine	TAT, in liver and brain	Fuller, 1970b
	TAT, brain norepinephrine	Black and Axelrod, 1970
E. theophylline	TAT	Fuller and Snoddy, 1970
	TAT	Kato, 1970b
	TAT, AIB transport	Scott et al., 1970

(TABLE 2 CONT.)

IV.	MISCELLANEOUS		
A.	Allylisopropyl-acetamide	TAT, tryptophan pyrrolase	Yuwiler et al., 1970a
B.	cyclic-nucleotides	TAT	Kato, 1970b
		TAT	Linarelli et al., 1970
		TAT, AIB transport	Scott et al., 1970
C.	oral contra-ceptives	TAT, plasma tyrosine	Rose and Cramp, 1970
D.	pyridoxine	TAT	Black, 1970b
E.	pneumococcal infection	TAT	Herman, 1970
F.	quinolinic acid	TAT	Hardeland, 1970a,b

* Literature review for this summary was completed on September 1. Citations to reports published before 1970 may be found in the above papers.

TABLE 3

INDUCTION OF THE AMINO ACID TRANSPORT SYSTEM IN RESPONSE TO ENVIRONMENTAL STRESS: SUMMARY OF 1970 REPORTS ON THE REGULATION OF THE HEPATIC AMINO ACID TRANSPORT SYSTEM

System	Stimulus	Reference
I IN VIVO	surgical injury	Shihabi et al, 1970
	dibutyryl-cyclic-AMP glucagon theophylline	Scott et al, 1970
II IN VITRO		
perfused liver	cyclic-AMP dibutyryl-cyclic-AMP glucagon	Chambers et al, 1970 Chambers and Bass, 1970
liver slices	cyclic-AMP dibutyryl-cyclic-AMP epinephrine glucagon hydrocortisone acetate theophylline	Tews et al, 1970 Tews and Woodcock, 1970

ACKNOWLEDGMENTS

This study was done during the tenure of U.S.P.H.S. postdoctoral fellowship, 1 F2 CA 32836 (D.F.S.) and 1 F2 CA 43880 (F.R.B.), and was supported in part by grants from the National Cancer Institute, U.S.P.H.S. CA-07175 and CRTY-5002.

REFERENCES

Akedo, H. and H.N. Christensen. 1962. Transfer of amino acids across the intestine: a new model amino acid. J. Biol. Chem. 237: 113-117.

Aschoff, J. 1967. Adaptive cycles: their significance for defining environmental hazards. Intern. J. Biometeor. 11: 255-278.

Baril, E.F. and V.R. Potter. 1968. Systematic oscillations of amino acid transport in liver from rats adapted to controlled feeding schedules. J. Nutr. 95: 228-237.

Baril, E.F., V.R. Potter, and H.P. Morris. 1969. Amino acid transport in rat liver and Morris hepatomas: Effect of protein diet and hormones on the uptake of α-aminoisobutyric acid-^{14}C. Cancer Res. 29: 2101-2115.

Baril, E.F., M. Watanabe, and V.R. Potter. 1968. Amino acid uptake and tyrosine transaminase activity in liver and hepatomas from rats adapted to controlled feeding schedules, p. 417-430. In A. San Pietro, M.R. Lamborg and F.T. Kenney [ed], Regulatory mechanisms for protein synthesis in mammalian cells, Academic Press, New York.

Black, I.B. 1970a. Induction of hepatic tyrosine aminotransferase mediated by a cholinergic agent. Nature 225: 648.

Black, I.B. 1970b. Regulation of hepatic tyrosine transaminase *in vivo* through interaction of norepinephrine and pyridoxal phosphate co-factor. J. Pharmacol. Exp. Therap. 174: 283-289.

Black, I.B. and J. Axelrod. 1970a. Regulation of hepatic tyrosine transaminase by norepinephrine. Fed. Proc. 29, 736.

Black, I.B. and J. Axelrod. 1970b. Biphasic effect of norepinephrine in the regulation of hepatic tyrosine transaminase activity. Arch. Biochem. Biophys. 138: 614-619.

Black, I.B. and J. Axelrod. 1970c. The regulation of some biochemical circadian rhythms, p. 135-155. In G. Litwack [ed], Biochemical actions of hormones, Volume I, Academic Press, New York.

Blake, R.L. 1970a. Control of liver tyrosine aminotransferase expression. Enzyme regulatory studies on inbred strains and mutant mice. Biochem. Genetics 4: 215-235.

Blake, R.L. 1970b. Hydrocortisone induction of tyrosine aminotransferase activity in genetically obese and diabetic mice. Effects of a multiple dosage schedule. Biochem. Pharm. 19: 1508-1512.

Blake, R.L. 1970c. Flumethasone induction of liver tyrosine aminotransferase activity in inbred strains and obese mutant mice. Biochem. Pharmacol. 19: 1803-1815.

Breckenridge, B., M. 1970. Cyclic AMP and Drug Action. Ann. Rev. Pharmacol., 10: 19-34.

Chambers, J.W. and A.D. Bass. 1970. Effect of glucagon and cyclic 3',5' - adenosine monophosphate on hepatic uptake of amino acid. Fed. Proc. 29: 616.

Chambers, J.W., R.H. Georg, and A.I. Bass. 1965. Effect of hydrocortisone and insulin on uptake of α-aminoisobutyric acid by isolated perfused rat liver. Mol. Pharmacol. 1: 66-76.

Chambers, J.W., R.H. Georg, and A.D. Bass. 1970. Effect of glucagon, cyclic-3',5'-adenosine monophosphate and its dibutyryl derivative on amino acid uptake by the isolated perfused rat liver. Endocrinology 87: 366-370.

Christensen, H.N. and J.C. Jones. 1962. Amino acid transport models: Renal resorption and resistance to metabolic attack. J. Biol. Chem. 237: 1203-1206.

Civen, M., C.B. Brown, and D.K. Granner. 1970. Biosynthetic control of diurnal rhythm of tyrosine alpha-ketoglutarate transaminase activity in rat liver. Biochem. Biophys. Res. Comm. 39: 290-295.

Cohn, C., D. Joseph, F. Lavin, W.J. Shoemaker, and R.J. Wurtman. 1970. Influence of feeding habits and adrenal cortex on diurnal rhythm of hepatic tyrosine transaminase activity. Proc. Soc. Exp. Biol. Med. 133: 460-462.

Diamondstone, T.I. 1966. Assay of tyrosine transaminase activity by conversion of p-hydroxyphenylpyruvate to p-hydroxybenzaldehyde. Anal. Biochem. 16: 395-401.

Fuller, R.W. 1970a. Daily variation in liver tryptophan, tryptophan pyrrolase and tyrosine transaminase in rats fed ad libitum or single daily meals. Proc. Soc. Exp. Biol. Med. 133: 620-622.

Fuller, R.W. 1970b. Differences in regulation of tyrosine aminotransferase in brain and liver. J. Neurochem. 17: 539-543.

Fuller, R.W. and H.D. Snoddy. 1970. Induction of tyrosine aminotransferase in rat liver by epinephrine and theophylline. Biochem. Pharmacol. 19:1518-1521.

Hardeland, R. 1970a. Tyrosin-Transaminase-Aktivität durch Chinolinsäure. Naturwissen. 57: 252-253.

Hardeland, R. 1970b. Stimulation of two adaptive rat liver enzymes by quinolinic acid as a function of time. Life Sci. Part II, 9: 901-906.

Herman, T.S. 1970. Effect of pneumococcal infection on hepatic tyrosine aminotransferase (TAT) in pregnant rats and their fetuses. Fed. Proc. 29: 776.

Holt, P.G., J.W. Sadlier, and I.T. Oliver. 1970. Multiple forms of tyrosine aminotransferase in rat liver, the solution of an inductive dilemma? Proc. Austral. Biochem. Soc. 3: 22.

Kato, K. 1970a. Selective repression of benzoate or tryptophan mediated induction of liver tyrosine aminotransferase by phentolamine in adrenalectomized animals. FEBS Letters 8: 316-318.

Kato, K. 1970b. Induction of liver tyrosine aminotransferase by theophylline and its repression by phentolamine and glucose in adrenalectomized rats. FEBS Letters 9: 105-107.

Kenney, F.T. 1967. Turnover of rat liver tyrosine transaminase: stabilization after inhibition of protein synthesis. Science 156: 525-528.

Kreutler, P.A. and S.A. Miller. 1970. Dietary regulation of hepatic tyrosine transaminase in the neonatal rat. Fed. Proc. 29: 364.

Levitan, I.B. and T.E. Webb. 1970a. Hydrocortisone-mediated changes in concentration of tyrosine transaminase in rat liver -- an immunochemical study. J. Mol. Biol. 48: 339-348.

Levitan, I.B. and T.E. Webb. 1970b. Posttranscriptional control in steroid-mediated induction of hepatic tyrosine transaminase. Science 167: 283-285.

Liberti, J.P., E.S. Longman, and R.S. Navon. 1970. Effects of hydrocortisone and growth hormone on tyrosine aminotransferase and tryptophan oxygenase levels in hypophysectomized rats. Endocrinol. 86: 1448-1450.

Linarelli, L.G., J.L. Weller, and W.H. Glinsmann. 1970. Stimulation of fetal rat liver tyrosine aminotransferase activity *in utero* by 3', 5'-cyclic nucleotides. Life Sci. Part II, 9: 535:539.

Mavrides, C. and E.A. Lane. 1970. The permissive role of cortisol in regulation of rat liver tyrosine aminotransferase. Canad. J. Biochem. 48: 13-19.

Nakajima, K., H. Matsutaka, and E. Ishikawa. 1970. Hormonal and dietary control of enzymes in the rat. 2. Effect of pancreatictomy and adrenalectomy on some liver enzymes unique to amino acid metabolism and gluconeogenesis. J. Biochem. 67: 779-787.

Noall, M.W., T.R. Riggs, L.M. Walker, and H.N. Christensen. 1957. Endocrine control of amino acid transfer. Science 126: 1002-1005.

Potter, V.R. 1969. How is an optimum environment defined? Environmental Research 2: 476-487.

Potter, V.R. 1970. Intracellular responses to environmental change: The quest for optimum environment. Environmental Research 3: 176-186.

Potter, V.R., E.F. Baril, M. Watanabe, and E.D. Whittle. 1968. Systematic oscillations in metabolic functions in liver from rats adapted to controlled feeding schedules. Fed. Proc. 27: 1238-1245.

Potter, V.R., R.D. Reynolds, M. Watanabe, H.C. Pitot, and H.P. Morris. 1970. Induction of a previously non-inducible enzyme in Morris hepatoma 9618A. Advances Enzyme Regul. 8: 299-310.

Riggs, T.R. 1964. Hormones and the transport of nutrients across cell membranes, p. 1-57. In G. Litwack and D. Kritchevsky [ed], Actions of hormones on molecular processes, Wiley, New York.

Riggs, T.R. 1970. Hormones and transport across cell membranes, p. 157-208. In G. Litwack [ed], Biochemical actions of hormones, Volume I, Academic Press, New York.

Riggs, T.R., R.B. Sanders, and H.K. Weindling. 1963. Hormonal modification of the distribution of 1-amino-cyclopentane carboxylic acid-1-^{14}C in the rat. Endocrinology 73: 789-792.

Rose, D.P. and D.G. Cramp. 1970. Reduction of plasma tyrosine by oral contraceptives and oestogens -- a possible consequence of tyrosine aminotransferase induction. Clin. Chim. Acta 29: 49-54.

Rose, C.M. and R.J. Wurtman. 1970. Daily rhythms in content and utilization of tyrosine in the whole mouse. Nature 226: 454-455.

Sadleir, J.W., P.G. Holt, and I.T. Oliver. 1970. Fractionation of rat liver tyrosine aminotransferase during course of purification. Further evidence for multiple forms of enzyme. FEBS Letters 6: 46-48.

Scott, D.F., R.D. Reynolds, H.C. Pitot, and V.R. Potter. 1970. Co-induction of the hepatic amino acid transport system and tyrosine aminotransferase by theophylline, glucagon and dibutyryl-cyclic AMP _in vivo_. Life Sci., Part 2, _9_: in press.

Sekeris, C.E., J. Niessing, and K.H. Seifart. 1970. Inhibition by α-amanitin of induction of tyrosine transaminase in rat liver by cortisol. FEBS Letters _9_: 103-104.

Shihabi, Z., H.F. Balegno, and O.W. Neuhaus. 1970. Injury-stimulated uptake of α-aminoisobutyric acid by rat liver. Biochem. Biophys. Res. Comm. _38_: 692-696.

Susten, S.S. and A. Kirksey. 1970. Influence of pyridoxine on tyrosine transaminase activity in maternal and fetal rat liver. J. Nutr. _100_: 369-374.

Tews, J.K. and N.A. Woodcock. 1970. Amino acid transport in rat liver slices as affected by glucagon, epinephrine or cyclic-AMP. Fed. Proc. _29_: 365.

Tews, J.K., N.A. Woodcock, and A.E. Harper. 1970. Stimulation of amino acid transport in rat liver slices by epinephrine, glucagon and adenosine 3',5'-monophosphate. J. Biol. Chem. _245_: 3026-3032.

Watanabe, M., V.R. Potter and H.C. Pitot. 1968. Systematic oscillations in tyrosine transaminase and other metabolic functions in liver of normal and adrenalectomized rats on controlled feeding schedules. J. Nutr. _95_: 207-227.

Watanabe, M., V.R. Potter, H.C. Pitot, and H.P. Morris. 1969. Systematic oscillations in metabolic activity in rat liver and hepatomas. Effect of hydrocortisone, glucagon, and adrenalectomy on diploid and other hepatoma lines. Cancer Res. _29_: 2085-2100.

Yuwiler, A., L. Wetterberg, and E. Geller. 1970a. Alterations in induction of tyrosine aminotransferase and tryptophan oxygenase by glucose pretreatment. Biochim. Biophys. Acta _208_: 428-433.

Yuwiler, A., L. Wetterberg, and E. Geller. 1970b. Tryptophan pyrrolase, tryptophan and tyrosine transaminase changes during allylisopropyl acetamide-induced porphyria in rat. Biochem. Pharm. _19_: 189-195.

TRANSPLANTABLE TUMORS AS AN INTERNAL STRESS ON HOST METABOLISM AND RESPONSIVENESS OF TUMORS TO APPLIED STRESS

Chung Wu and Jere M. Bauer

Departments of Biological Chemistry and of Internal Medicine, The University of Michigan Medical School, Ann Arbor, Michigan 48104

The growth of a malignant tumor within the body represents a type of metabolic stress for the host that is fundamentally different from that ordinarily encountered in adaptations made by the organism to a changing external environment. Although the tumor-host system is still dependent on the environment for its ultimate source of energy and essential nutrients, the malignant tumor and the host now must exist in a metabolic relationship, one with the other, since they share a common metabolic pool. Perhaps, the most gross, yet most valid, observation in cancer biology is that unrestrained growth of a malignant tumor will lead ultimately to death of the host. Although the precise cause of death from cancer is difficult to ascertain, the common observation of generalized wasting and loss of tissue substance seen in the host with advanced cancer is indicative of disrupted metabolic processes in the tissues of the host.

On the other hand, it appears equally cogent to study biochemical responses of tumors to metabolic stress. During the induction and final expression of malignancy, essential biochemical alterations in cell metabolism must have occurred. Whether there is a common biochemical denominator in all forms of malignancy is at present conceptual. In recent years, the introduction by Morris (1966) of malignant rat liver tumors with normal or nearly normal karyotypes (Nowell and Morris, 1969) has made such studies on biochemical responsiveness meaningful. However, these tumors show a constellation of enzyme patterns (Wu, 1967) not only different from one another, but different from that of normal liver at various stages of development. By studying their responses to environmental challenge, we hope to understand some of the regulatory mechanisms operating in malignancy, and to indicate areas of similarity between oncogenetic progression and ontogenetic development.

The experiments to be described were conceived along two lines. First, by considering the tumor as an internal stress, we investigate the biochemical responses of the host to this stress. Second, by subjecting the host-tumor system to an applied stress, we examine the biochemical responses of the host and the tumor to the administered stress.

These experiments have been done with several lines of transplantable tumors in the rat. Some of these tumors, such as the Walker carcinoma 256 and Novikoff hepatoma, have been used in experiments for several decades, and they do not resemble in any way their tissues of origin. These tumors always have rapid growth rates and kill the host in a matter of a few weeks. Other tumors, such as the Morris Hepatoma 7800 and 9618A, have been available for less than a decade, and they are similar to liver in certain morphologic, cytogenetic, and biochemical properties. They grow slowly, and it takes several months for them to kill the host.

We have used the fast growing, undifferentiated, grossly aneuploid tumors to study their metabolic effects on the host, because these tumors are likely to produce greater effects in shorter periods of time. On the other hand, we have used the slow growing, well differentiated, diploid and aneuploid liver tumors to study their responses to the administered metabolic stress because of their resemblance to normal liver.

RESPONSE OF HOST TO TUMOR

Decrease in appetite and loss of body weight are common symptoms of the tumor-bearing host. Because of this and the great demands of the growing tumor for nitrogen, there is a progressive loss of nitrogenous constituents in the tissues of the host. This deficiency intensifies as the tumor grows, and it often manifests itself in the increased excretion of urinary end-products. At the same time, the activity of many enzymes in the host tissues also undergoes significant changes.

Urinary Excretion

The tumor used in these experiments was Walker carcinoma 256. In order to equalize the food intake which could affect the level of excretion, we pair-fed a control with each tumor-bearing rat. Figure 1 shows the excretion of uric acid by the control and tumor-bearing rats with increasing weight of the tumor (Wu and Bauer, 1962). Uric acid is a degradative product of purine metabolism in the rat. We can see that the excretion of uric acid by the tumor-bearing rat increased progressively with the increasing weight of the tumor, while the excretion by the control rat remained constant. These results indicate that the tumor imposes on the host a stress that increases the degradative activities of the host.

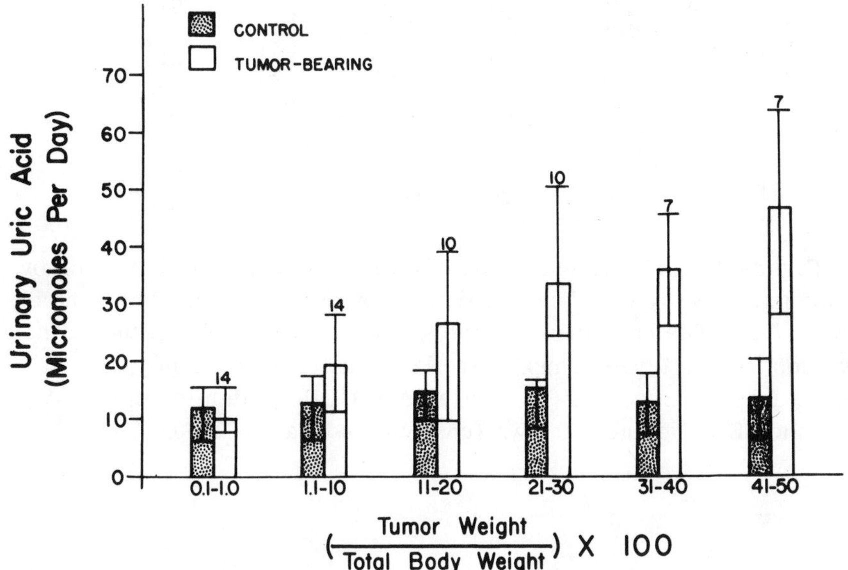

Figure 1. Urinary excretion of uric acid by male Sprague-Dawley rats bearing Walker carcinoma 256 and their pair-fed controls. Height of bar indicates the mean value for each group. The vertical line through each bar indicates the range of values, and the numerical figure represents the number of animals used. From Wu and Bauer (1962).

Figure 2 shows the excretion of dialyzable, bound amino acids by rats bearing Walker carcinoma 256 (Wu and Bauer, 1960a). Presumably, these amino acids were excreted as simple peptides or as conjugates. We see that the excretion was greatly increased in tumor-bearing rats, and generally, the increase was greater when the tumor was larger. On the other hand, there was no increase in the excretion of the free amino acids by tumor-bearing rats.

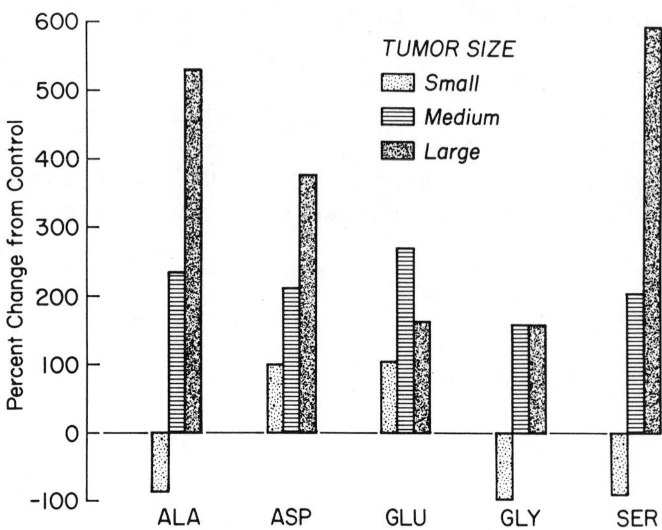

Figure 2. Urinary excretion of dialyzable bound amino acids by male Sprague-Dawley rats bearing Walker carcinoma 256. The controls were pair-fed. The numbers of urine samples used to obtain the averaged values for the three groups were 12 (small), 10 (medium), and 3 (large). ALA, alanine; ASP, aspartic acid; GLU, glutamic acid; GLY, glycine; and SER, serine. Drawn from Wu and Bauer (1960a).

The excretion of uric acid and bound amino acids is an example of the gradual increase in the excretion of urinary end products with tumor growth. We show next a different kind of aberration in urinary excretion. Figure 3 shows the excretion of creatine and creatinine by a rat bearing Walker carcinoma 256. Creatine, a normal constituent of skeletal muscle, appears in trace amounts in normal urine, but creatinine is excreted normally as one of the more abundant nitrogenous end products in the urine. During the early periods of tumor growth, there was no apparent disparity in the excretion of either creatine or creatinine by the tumor-bearing rat, when compared with the excretion of the pair-fed control. Near the terminal stage of tumor growth, however, the excretion of creatine abruptly rose to a level a few hundred times the level excreted 24 hours earlier. Although the onset of creatinuria varied in time from one host to another, this steep rise in creatine excretion invariably signaled the approaching death of the host.

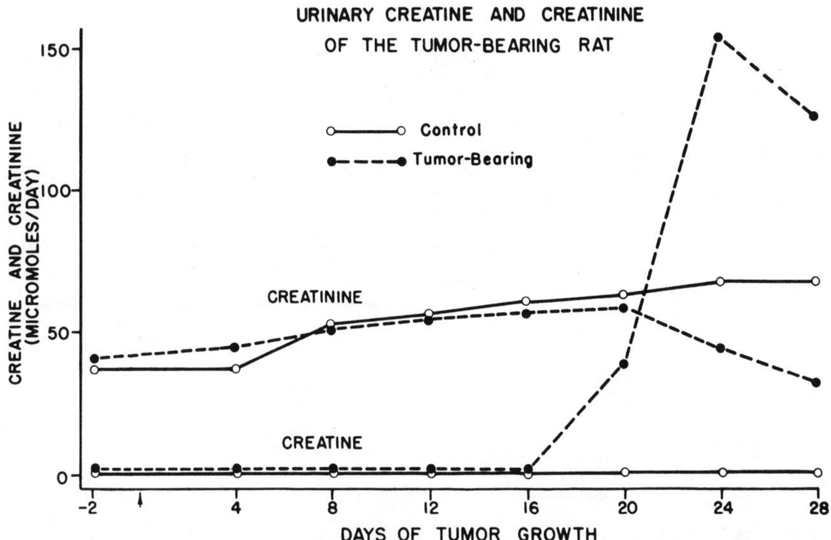

Figure 3. Urinary excretion of creatine and creatinine by a male Sprague-Dawley rat bearing Walker carcinoma 256 and its pair-fed control during the entire period of tumor growth. Creatine and creatinine were determined according to Taussky (1954).

Liver Enzymes

The activity of many enzymes in host tissues, notably liver, is affected by tumor growth. Generally, the effect becomes greater with longer periods of tumor growth. In these experiments we used rats bearing Novikoff hepatoma, a fast growing, undifferentiated tumor. The results depicted in Fig. 4 show that there was no apparent decrease in food intake until the eighth day after transplantation of the tumor, but changes in activity of some enzymes in the host liver took place much sooner (Wu and Homburger, 1969). This means that changes in the rate of production or degradation or both of certain enzymes precede loss of the appetite for food.

Of the five enzymes studied, two showed an increase in activity, and three showed a decrease. The increase in glucose 6-phosphate dehydrogenase activity was striking. By the fourth day after transplantation of the tumor which was still not palpable, the increase was already 35% above the control level. The greatest increase in activity occurred between the second and fourth days, but it continued to the sixth day, after which time no appreciable increase ensued.

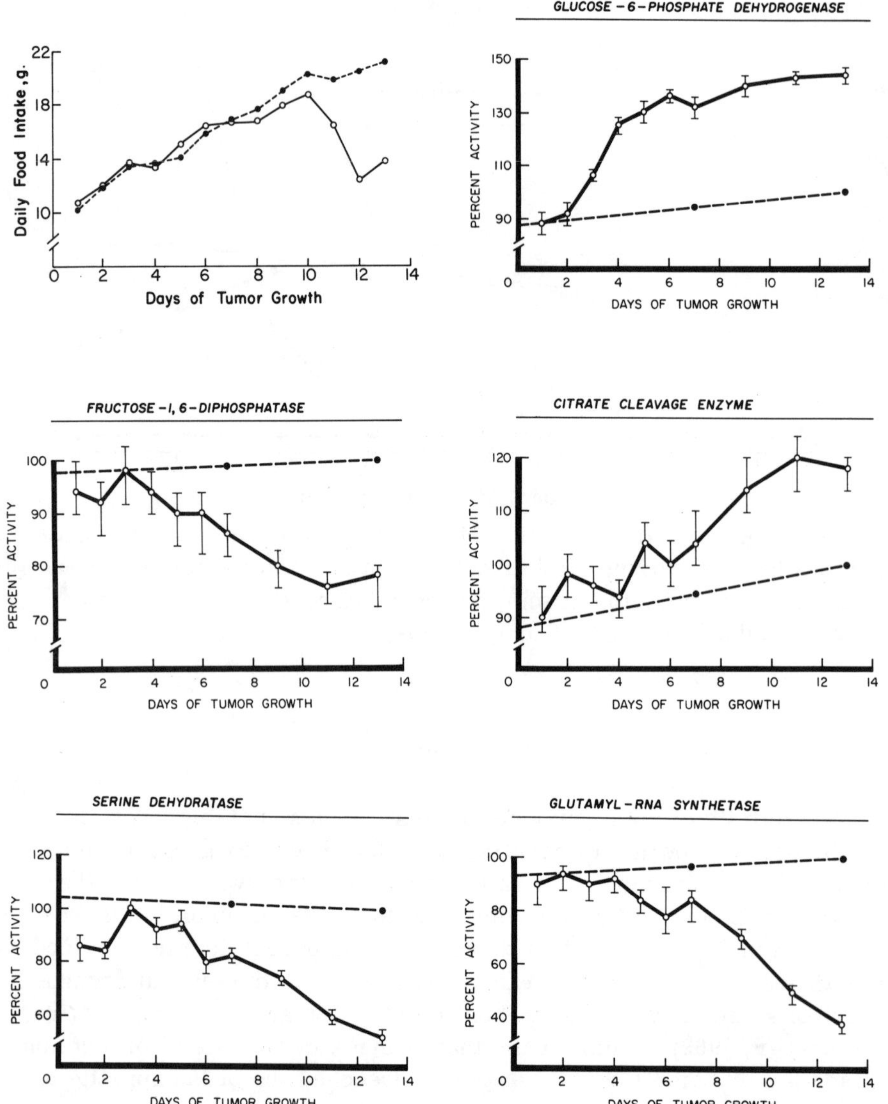

Figure 4. Daily food intake of normal Sprague-Dawley (●) and tumor-bearing (O) rats of the same sex and age (upper left), and changes in the activity of five enzymes in the host liver with growth of Novikoff hepatoma. The activity of each enzyme in liver of normal rats killed at the end of the experimental period is chosen as 100%. (●) shows activity in normal liver, and (O), activity in host liver. The vertical line indicates the range. From Wu and Homburger (1969).

The patterns of change during tumor growth for citrate cleavage enzyme and for fructose 1,6-diphosphatase are similar, although the changes took place in opposite directions. The tumor growth caused a gradual change in activity of the two enzymes in the first 6 days, after which time the change intensified with growth of the tumor. By the end of the 11th day, no further increase or decrease was observed.

On the other hand, there was no appreciable change in activity of serine dehydratase and glutamyl-RNA synthetase in the first 5 days following inoculation of the tumor. After that, the activity declined rapidly with days of tumor growth. The decrease was still continuing at the 13th day when the experiment terminated.

These results show that growth of Novikoff hepatoma has brought about different patterns of change for different enzymes in the host liver. Some enzymes are more responsive to tumor growth than others. Of course, there are also enzymes unresponsive to tumor growth. The different patterns of change seen in these experiments reflect this difference in responsiveness.

Although tumor growth can affect the activity of many enzymes in the host liver, it may not have similar effects on the enzymes in the host brain. Table 1 shows a comparison of the response of glutamine synthetase in liver and in brain to growth of a number of tumors (Wu et al., 1965). Evidently, glutamine synthetase in brain did not respond to tumor growth, while the enzyme in liver did.

Kidney Transamidinase

This enzyme catalyzes the formation of guanidoacetic acid from arginine and glycine, and resides in the kidney. Using the pair-feeding technique, we studied changes of the enzyme activity in the host kidney at different stages of tumor growth (Wu and Bauer, 1960b). Table 2 shows that, during growth of Walker carcinoma 256, the enzyme activity in the host kidney declined steadily. The decline intensified as the size of the tumor increased. This table also shows the concentration of guanidoacetic acid in the host liver. Although transamidinase activity of kidney declined with tumor growth, the concentration of guanidoacetic acid in liver increased. This accumulation suggests that the host liver has retained its capacity to concentrate guanidoacetic acid synthesized in the host kidney, but that it has shown an impairment in its ability to utilize guanidoacetic acid for creatine synthesis. This impairment in creatine synthesis in the face of a massive creatinuria shown earlier points to a deranged

TABLE 1

GLUTAMINE SYNTHETASE ACTIVITY IN
LIVER AND BRAIN OF TUMOR-BEARING RATS

Tumor in rats	Activity[a]	
	Liver	Brain
None(Sprague-Dawley)(6)[b]	204 ± 27[c]	69.2 ± 4.8
Walker (6)	134 ± 10	
Murphy-Sturm (8)	111 ± 19	67.6 ± 10.8
Jensen (6)	84.8 ± 12.0	59.6 ± 9.2
Guerin (6)	112 ± 16	
Novikoff (6)	157 ± 23	68.0 ± 8.8
None (ACI) (6)	201 ± 37	44.0 ± 4.8
Morris 3924A (6)	274 ± 56	47.6 ± 3.6
None (Buffalo) (10)	202 ± 17	68.8 ± 6.4
Morris 5123B (9)	268 ± 74	66.0 ± 5.2

[a]Glutamine synthetase activity is expressed as units/g. One unit of activity is the amount of enzyme that will catalyze the formation of 1 μmole of γ-glutamylhydroxamate in 1 hr at 37°.

[b]In this table and the others, the number in parentheses indicates the number of animals used.

[c]In this table and the others, the value following the ± sign is the standard deviation. Where the experiment used fewer than 3 animals, the individual observations are given instead.

From Wu et al. (1965).

creatine metabolism in the muscle of the host.

We have seen from the results presented above that tumor growth affects host metabolism in many ways. This effect is manifested in altered patterns of urinary excretion and in changes in activity of a variety of enzymes in liver and kidney. The magnitude of these changes is clearly a function of the tumor size. When the tumor is small, usually no alteration is noticeable. Presumably, during the initial stages of tumor growth, the defense systems of the host are still capable of

TABLE 2

KIDNEY TRANSAMIDINASE ACTIVITY AND LIVER GUANIDOACETIC ACID OF RATS BEARING WALKER CARCINOMA 256

Rats	Tumor size[a]		
	Small	Medium	Large
		Transamidinase activity[b]	
Control	60.0 ± 5.9(3)	52.3 ± 8.5(6)	42.4 ±5.4 (8)
Tumor-bearing	49.3 ±19.7	36.5 ±14.6	12.3 ±2.8
		Guanidoacetic acid[c]	
Control	1.34± 0.34(5)	1.32± 0.20(5)	1.38±0.25(5)
Tumor-bearing	1.67± 0.32	1.74± 0.23	2.09±0.42

[a]Small, tumors weighing less than 10% of total body weight; medium, 10-30%; large, more than 30%.

[b]The enzyme activity is expressed as μmoles of guanidoacetic acid formed/g kidney (dry weight)/hr at 37°.

[c]Micromoles/g liver (dry weight).

From Wu and Bauer (1960b).

accommodating the stress and thereby show no sign of deterioration. However, as the tumor growth continues, the host system gradually yields to the increasing stress with changes in cellular function.

RESPONSE OF TUMOR AND HOST TO ADMINISTERED STRESS

When the host-tumor system is subject to a metabolic modulation, the host and the tumor presumably are exposed to the same stress. However, the responsiveness to stress of several enzymes in the host tissue differs from that in the tumor. We shall first examine the aberrations under a variety of metabolic stress and then try to understand the meaning of these aberrations in relation to malignancy.

Hormonal Treatment

Cortisol. Glutamine synthetase in adult rat liver cannot be readily induced or repressed by metabolic manipulations. Ammonium salts and glutamate, when administered *in vivo*, did not increase the enzyme activity in rat liver, nor did glutamine decrease it (Wu, 1964). Although cortisol injection moderately raised the enzyme activity in liver of 2-week old rats, it did not do so in adult liver (Wu, 1964). However, the enzyme in a number of hepatomas was highly responsive to cortisol treatment. Table 3 shows the elevation of glutamine synthetase activity by cortisol in four hepatomas (Wu and Morris, 1970). Generally, the hormone raised the enzyme activity 2- to 3-fold in the hepatomas after 2 days of treatment. In contrast, the hormone did not affect the enzyme activity in the host livers.

TABLE 3

CORTISOL INDUCTION OF GLUTAMINE SYNTHETASE IN HEPATOMAS

Cortisol suspended in sesame oil was given intramuscularly, 50 μg/g body weight daily for 2 days.

Rats bearing	Activity	
	Host liver	Hepatoma
(a) Hepatoma 7800 (X-XIII)[a]		
Control (13)	228±28	277 ± 82
Treated (7)	214±26	656 ±128
(b) Hepatoma 7787 (IX)		
Control (5)	183±66	66.0± 14.0
Treated (3)	205±11	136 ± 7
(c) Hepatoma 9618A (VI)		
Control (3)	301±12	124 ± 20
Treated (3)	284±16	395 ±220
(d) Hepatoma 8999 (IX)		
Control (4)	183±24	695 ±385
Treated (3)	193±19	1424 ±854

[a]The roman numerals indicate the generation number of the tumor.

From Wu and Morris (1970).

Table 4 shows further that cortisol raised glutamine synthetase activity some 11-fold in hepatoma 7800, but none in the host liver (Wu and Morris, 1970). (Comparison of these results with those of Hepatoma 7800 in Table 3 will show that the tumor generation has affected both the basal activity of the enzyme and its inducibility.) On the other hand, there was a moderate response to cortisol by the enzyme in the liver of 3-week old rats. Hence, with respect to glutamine synthetase, the hepatomas responded in a way similar to the weanling rat liver. Moreover, adrenalectomy lowered the enzyme activity by 50% in the weanling rat liver. Although this lowering lagged behind the 80% reduction seen in the hepatoma, it stands in sharp contrast to the unresponsiveness of the enzyme in the adult host liver. This observation serves as another example of the similarity in response by the enzyme in juvenile liver with that in hepatoma.

Actinomycin D completely blocked the induction of glutamine synthetase by cortisol, when given at the time of cortisol treatment (Table 4). The result suggests that the enzyme induction by cortisol depends on continuous RNA synthesis. However, the antibiotic affected neither the enzyme activity in the host liver nor the basal activity in the hepatoma.

Thyroxine. Like cortisol, thyroxine also increased glutamine synthetase activity in the liver of 2-week old rats (Wu, 1964). Table 5 shows that the hormone had no effect on the enzyme activity in the liver of rats bearing Hepatoma 7800. On the other hand, thyroxine raised the enzyme activity about 9-fold in Hepatoma 7800 in 7 days. Obviously, the enzyme in the tumor responded to thyroxine treatment, but the enzyme in the host liver did not.

Furthermore, we also found that at least two other enzymes in the hepatoma were affected by thyroxine injection. Table 6 shows the induction of argininosuccinate synthetase (Wu et al., 1971) and glutamine aminotransferase (Wu and Morris, 1970) by thyroxine in Hepatoma 7800 grown in thyroidectomized rats. The prevention of the induction by actinomycin D again suggests a dependence of the induction on continuous RNA synthesis. On the other hand, neither the hormone nor a combination of the hormone and the antibiotic exerted any effect on these two enzymes in the host liver.

Glucagon. The hormone lowered glutamine synthetase activity in both the host liver and Hepatoma 8999. Table 7 shows the results. In 3 days of repeated injections, the hormone reduced the enzyme activity to 35% of the basal level in the host liver, but it decreased the enzyme activity to 4% of the basal value in the hepatoma. The precipitous loss

TABLE 4

EFFECT OF CORTISOL, ACTINOMYCIN D AND ADRENALECTOMY ON GLUTAMINE SYNTHETASE OF RAT LIVER AND HEPATOMA 7800

Male Buffalo rats bearing Hepatoma 7800 (generation XLIII) received cortisol as prescribed in the preceding table. Actinomycin D, 0.2 μg/g body weight, was injected intraperitoneally daily alone or at the time of cortisol treatment. Bilateral adrenalectomy was done a few days after tumor inoculation and 60 days before sacrifice. Male adult and young Sprague-Dawley rats also received cortisol. Young rats were 3 weeks old at the beginning of the experiment; bilaterally adrenalectomized young rats were used 7 days after the surgery.

Treatment	Activity			
	Weanling liver	Adult liver	Host liver	Hepatoma
None	180 ± 18(14)	250 ± 34 (6)	243 ± 34 (4)	69.6 ± 17.2 (4)
Cortisol, 2 days	257 ± 8 (3)[a]	251 ± 37 (4)[b]	246 (238, 254)	796 (756, 836)
Actinomycin, 2 days			300 ± 50 (4)	54.4 ± 13.4 (3)
Cortisol + actinomycin, 2 days			235 ± 35 (5)	75.6 ± 11.2 (4)
Adrenalectomy	90.8 ± 12.8 (9)		254 ± 40 (4)	15.2 ± 4.0 (3)

[a] After 3 days of treatment.

[b] After 4 days of treatment.

From Wu and Morris (1970).

TABLE 5

THYROXINE INDUCTION OF GLUTAMINE SYNTHETASE IN THE HEPATOMA 7800

Male Buffalo rats bearing Hepatoma 7800 (generation XLIII) received subcutaneous injection of L-thyroxine, 1 μg/g body weight daily for 7 days.

Treatment	Activity	
	Host liver	Hepatoma
None (4)	243 ± 34	69.6 ± 17.2
Thyroxine (3)	290 ± 10	624 ± 48

TABLE 6

EFFECT OF ACTINOMYCIN D ON THYROXINE INDUCTION OF ARGININOSUCCINATE SYNTHETASE AND GLUTAMINE AMINOTRANSFERASE IN HEPATOMA 7800

Male Buffalo rats bearing Hepatoma 7800 (generation XLV) were bilaterally thyroidectomized a few days after inoculation of the tumor. Fifty days later, thyroxine was given as described in the preceding table. Actinomycin D was injected intraperitoneally, 0.2 μg/g body weight daily, on the last 2 days of thyroxine treatment.

Treatment	Activity[a]	
	Host liver	Hepatoma
(a) Argininosuccinate synthetase		
None (3)	83.4(75.9, 90.9)	101 ±23
Thyroxine (3)	79.1(73.9, 84.2)	312 ±13
Thyroxine + actinomycin (3)	68.7±20.5	168 ±12
(b) Glutamine aminotransferase		
None (3)	24.3± 5.5	17.1± 6.6
Thyroxine (3)	24.6± 4.5	32.0± 1.9
Thyroxine + actinomycin (3)	24.0± 1.9	18.2± 0.4

[a]The synthetase activity is expressed as μmoles of urea formed/g/hr at 37°; the transferase activity, as μmoles of phenylpyruvate converted/g/hr at 25°. The synthetase activity was assayed according to Schimke (1962); the transferase activity, according to Kupchik and Knox (1970).

From Wu and Morris (1970), and Wu et al. (1971).

TABLE 7

REPRESSION OF GLUTAMINE SYNTHETASE BY GLUCAGON IN HOST LIVER AND HEPATOMA 8999

Male Buffalo rats bearing Hepatoma 8999 (generation IX) received glucagon, 1.5 μg/g body weight, subcutaneously every 12 hours for 3 days.

Treatment	Activity	
	Host liver	Hepatoma
None (4)	183 ± 24	695 ± 384
Glucagon (3)	52.0 ± 13.8	28.5 ± 11.6

of the enzyme in the hepatoma was indeed phenomenal. Of course, this effect of glucagon can be explained by a decrease in the rate of synthesis, or an increase in the rate of degradation, or both. From what we shall see later in the experiment with cycloheximide, it becomes obvious that, if glucagon is to regulate the enzyme synthesis alone, it would have to exert a complete block in order to bring the enzyme activity down to 4% in 3 days. Perhaps, the effect of glucagon comes in part from an acceleration of enzyme degradation or from an inactivation.

The three hormones used in the present study have been shown to induce many enzymes in fetal, neonatal, and adult rat livers, such as cortisol on tyrosine aminotransferase (Lin and Knox, 1957; Kenney, 1962) and tryptophan pyrrolase (Civen and Knox, 1959; Feigelson and Greengard, 1963), thyroxine on argininosuccinate synthetase (Freeland and Sodikoff, 1962), glucose 6-phosphatase and $NADPH_2$ dehydrogenase (Greengard and Dewey, 1968), triiodothyronine on α-glycerophosphate dehydrogenase (Hunt et al., 1970), and glucagon on tyrosine aminotransferase (Kenney et al., 1968), phosphopyruvate carboxylase (Yeung and Oliver, 1968), threonine dehydratase and ornithine aminotransferase (Peraino and Pitot, 1964). We found that, although cortisol and thyroxine were incapable of inducing the enzymes we studied in liver of adult rats, they were capable of doing so in the hepatomas. The two hormones also induced glutamine synthetase in liver of suckling rats (Wu, 1964). We showed the diminution of glutamine synthetase activity by glucagon to be highly significant in the adult liver but even more so in the hepatoma.

Surgical Manipulations

We may consider surgical manipulations as an extension of hormonal treatment, since the surgery was concerned with hormone-producing glands. We have shown earlier in Table 4 that adrenalectomy lowered glutamine synthetase activity to 20% of the control value in Hepatoma 7800, but it did not affect the enzyme activity in the host liver. The results suggest that the maintenance of the enzyme activity in the hepatoma requires the constant presence of corticosteroids, although no such requirement is evident in the host liver. Apparently, the tumor can be more dependent on its physiologic environment for its biochemical activities than the host tissue. In Table 8, we show that thyroidectomy greatly reduced iodide peroxidase activity of a thyroid tumor. This enzyme is concerned with the oxidation and activation of iodide in the biosynthesis of iodotyrosines. Hence, the ability of this thyroid tumor to carry out its biologic function is also contingent upon the hormonal constituents of its environment.

TABLE 8

EFFECT OF THYROIDECTOMY ON IODIDE PEROXIDASE ACTIVITY IN THYROID TUMOR

Bilaterally thyroidectomized male Fischer rats were inoculated with a fairly well differentiated follicular carcinoma of the thyroid several days after the surgery, and were used for the enzyme assay about 7 weeks after the inoculation. Intact rats bearing the same tumor served as the control.

Rats	Activity[a]	
	Host thyroid	Tumor
Intact (12)	329 ± 104	446 ± 203
Thyroidectomized (3)		123 ± 31

[a]The enzyme activity assayed according to Mahoney and Igo (1966) is expressed as μmoles of iodide oxidized/g/hr at 25°.

Effect of Metabolic Inhibitors

We have shown the inhibitory effect of actinomycin D on the inductive increase in glutamine synthetase, argininosuccinate synthetase, and glutamine aminotransferase by cortisol or thyroxine. In each case, the enzyme in the hepatoma was responsive to the antibiotics, but that in the host liver was not. In Table 9, we show how actinomycin D and cycloheximide blocked the cortisol-induced increase in arginase activity in Hepatoma 8999 (Wu et al., 1971). There was a nearly 10-fold increase in the enzyme activity in the tumor after 2 days of cortisol treatment. When each of the antibiotics was given on the second day of the hormone injection, it inhibited most of the increase. On the contrary, arginase in the host liver did not respond to cortisol, nor to the antibiotics.

TABLE 9

EFFECT OF METABOLIC INHIBITORS ON CORTISOL-INDUCED ARGINASE ACTIVITY IN HEPATOMA 8999

Male Buffalo rats bearing Hepatoma 8999 (generation VI) received cortisol, 50 μg/g body weight daily for 2 days. Actinomycin D, 1 μg/g body weight, or cycloheximide, 5 μg/g body weight, was injected intraperitoneally for 1 day--on the second day of cortisol treatment.

Treatment	Activity[a]	
	Host liver	Hepatoma
None (5)	98.7 ± 21.6	0.65 ± 0.19
Cortisol (3)	115 ± 19	6.20 ± 1.49
Cortisol + actinomycin (3)	117 ± 10	1.75 ± 0.21
Cortisol + cycloheximide (3)	115 ± 28	3.44 ± 0.89

[a]Arginase activity assayed according to Schimke (1962) is expressed as mmoles of urea formed/g/hr at 37°.

From Wu et al. (1971).

Table 10 shows the response of arginase in the liver of 3-week old rats to actinomycin D and cycloheximide. Apparently, the enzyme in the juvenile rat liver responded to the metabolic inhibitors, to some degree, like the enzyme in Hepatoma 8999, but unlike that in the adult host liver (Table 9).

TABLE 10

EFFECT OF METABOLIC INHIBITORS ON ARGINASE ACTIVITY OF WEANLING RAT LIVER

Male 3-week old Sprague-Dawley rats weighing 50-60 g received a single injection of actinomycin D, 2 μg/g body weight, or cycloheximide, 5 μg/g body weight, and were used 12 hours later for the enzyme assay.

Treatment	Activity
None (14)	58.9 ± 2.7
Actinomycin (4)	46.7 ± 2.3
Cycloheximide (4)	41.6 ± 0.2

Table 11 shows another example of over-responsiveness by an enzyme in hepatoma. In 12 to 16 hours following an intraperitoneal injection of cycloheximide, glutamine synthetase activity in Hepatoma 7800 declined to less than one-half of its control level, while the activity in the host liver did not change.

Comparison of a partially purified preparation of glutamine synthetase from normal adult rat liver with that from Hepatoma 7800 shows that the two proteins were not different. Table 12 summarizes the results. Although additional experiments with highly purified enzyme preparations are still needed, we are inclined to think that the difference in responsiveness presented in this study reflects a difference in cellular regulatory mechanisms in these two tissues rather than a difference in the properties of the enzyme.

In the experiments described above, where a comparison of responsiveness between the adult host liver and the hepatoma has been attempted, we have shown the activity of several enzymes in different

TABLE 11

EFFECT OF CYCLOHEXIMIDE ON GLUTAMINE SYNTHETASE ACTIVITY IN HEPATOMA 7800

Male Buffalo rats bearing Hepatoma 7800 (generation XXI) received an intraperitoneal injection of cycloheximide, 5 μg/g body weight. At the end of different time intervals after the injection, the animals were used for the enzyme assay.

Hours after injection	Activity	
	Host liver	Hepatoma
0 (13)	228 ± 28	277 ± 82
8 (2)	225 (223, 228)	257 (256, 258)
12 (3)	241 ± 6	112 ± 16
16 (3)	228 ± 9	101 ± 3

TABLE 12

SOME PROPERTIES OF GLUTAMINE SYNTHETASE FROM LIVER AND HEPATOMA 7800

Property	Source of enzyme	
	Normal liver[a]	Hepatoma
$K_m \times 10^4$ $\underline{M}$:		
ATP	2.0	2.3
NH_2OH	1.1	1.4
Glutamate	22	20
SH requirement	Yes	Yes
Inhibition by arsenite-dithiol	Yes	Yes

[a]From Wu (1965).

lines of tumors under a variety of stresses. These stresses include hormonal treatment, surgical removal of certain endocrine glands, and administration of metabolic inhibitors. The results of these experiments bring forth two salient points. First, certain enzymes in the tumors display an over-responsiveness to metabolic modulations. Second, in some respects, the tumors are more dependent than the host on the environment to maintain their metabolic activities.

SOME CONCLUDING OBSERVATIONS

Investigators in several laboratories (Morris et al., 1964; Potter et al., 1969; Sheid and Roth, 1965; Siperstein et al., 1966; and Weber et al., 1965) have demonstrated the lack of responsiveness to metabolic modulations of enzymes in many hepatomas. These observations are in contrast, but not in contradiction, to our findings presented here. Indeed, we would consider the lack of responsiveness and over-responsiveness as supplementary to each other, for they both reveal derangements in control mechanisms in malignancy. Before we can explain malignancy in molecular terms, we must take into consideration the fact that these two kinds of responsiveness do exist in the cancer cell.

We should like to consider over-responsiveness as a result of the re-acquisition or reactivation during carcinogenesis of certain factors of the regulatory machinery. We further suggest that these factors are operative, at least to some extent, in the cells during the early periods of ontogeny, but they are later suppressed as the development of the organism proceeds. In cancer cells, however, the suppression is lifted thus giving rise to certain biochemical properties akin to those present in cells at the early stages of development. Moreover, if the enzyme profile in the liver of developing rats is less mature than that of adult rats, the enzyme profile in the hepatoma seems never to have reached maturity. The contrast in responsiveness between the adult liver and the hepatoma, and the similarity in responsiveness between the developing liver and the hepatoma are in accord with this idea. Potter (1969) has discussed the importance of studies on fetal and neonatal tissues with reference to comparison with tumors.

In the parlance of cancer research, autonomy is usually considered as a sine qua non of malignancy. Although autonomy is difficult to define, it exists, no doubt, at different levels. We have shown the dependence of the activity of certain enzymes of tumors on their environment. In this respect, some malignant tumors not only have not attained autonomy, but definitely rely on their environment to determine their intrinsic properties and are more responsive to metabolic modulations than the normal tissue of origin.

The results presented in this study have emphasized the biochemical responses of the host to the stress of the tumor and the biochemical responses of the tumor to administered stress. Both the internal and the external stresses have played an important role in manipulating the metabolic activities of the affected cells. Perhaps, the biochemical responses seen in the host represent the inability of the host cells to accommodate the metabolic stress imposed by the tumor. On the other hand, the over-responsiveness seen in the tumor indicates that the tumor does not read the signal, in most cases, in the same way as normal liver. Presumably, the regulatory setup in the tumor is less elaborate than that in adult normal liver. As a result, the tumor cell seems to have a greater latitude in responsiveness to metabolic modulations. For a cancer cell, even though it may appear "autonomous," must deal with its environment for survival.

ACKNOWLEDGEMENTS

Previously unreported work was supported in part by a research grant, AM-07319, from the National Institute of Arthritis and Metabolic Diseases, U. S. Public Health Service.

REFERENCES

Civen, M. and W.E. Knox. 1959. The independence of hydrocortisone and tryptophan inductions of tryptophan pyrrolase. J. Biol. Chem. 234: 1787-1790.

Feigelson, P. and O. Greengard. 1963. Immunochemical evidence for increased titers of liver tryptophan pyrrolase during substrate and hormonal enzyme induction. J. Biol. Chem. 237: 3714-3717.

Freeland, R.A. and C.H. Sodikoff. 1962. Effect of diets and hormones on two urea cycle enzymes. Proc. Soc. Exper. Biol. Med. 109: 394-396.

Greengard, O. and H.K. Dewey. 1968. The developmental formation of liver glucose 6-phosphatase and reduced nicotinamide adenine dinucleotide phosphate dehydrogenase in fetal rats treated with thyroxine. J. Biol. Chem. 243: 2745-2749.

Hunt, S.M., M. Osnos, and R.S. Rivlin. 1970. Thyroid hormone regulation of mitochondrial α-glycerophosphate dehydrogenase in liver and hepatoma. Cancer Res. 30: 1764-1768.

Kenney, F.T. 1962. Induction of tyrosine-α-ketoglutarate transaminase in rat liver. III. Immunochemical analysis. J. Biol. Chem. 237: 1611-1614.

Kenney, F.T., J.R. Reel, C.B. Hager, and J.L. Wittliff. 1968. Hormonal induction and repression, p. 119-142. In A. San Pietro M.R. Lamborg, and F.T. Kenney [ed], Regulatory mechanisms for protein synthesis in mammalian cells, Academic Press, New York.

Kupchik, H.Z. and W.E. Knox. 1970. Assays of glutamine and its aminotransferase with the enol-borate of phenylpyruvate. Arch. Biochem. Biophys. 136: 178-186.

Lin, E.C.C. and W.E. Knox. 1957. Adaptation of the rat liver tyrosine-α-ketoglutarate transaminase. Biochim. Biophys. Acta 26: 85-88.

Mahoney, C.P. and R.P. Igo. 1966. Studies of the biosynthesis of thyroxine. II. Solubilization and characterization of an iodide peroxidase from thyroid tissue. Biochim. Biophys. Acta 113: 507-519.

Morris, H.P. 1966. The development of hepatomas of different growth rate; with comments on their biology and biochemistry. Gann Monograph 1: 1-10.

Morris, H.P., H.M. Dyer, B.P. Wagner, H. Miyaji, and M. Recheigl, Jr. 1964. Some aspects of the development, biology and biochemistry of rat hepatomas of different growth rate. Advances in Enzyme Regulation 2: 321-333.

Nowell, P.C. and H.P. Morris. 1969. Chromosomes of "minimal deviation" hepatomas: A further report on diploid tumors. Cancer Res. 29: 969-970.

Peraino, C. and H.C. Pitot. 1964. Studies on the induction and repression of enzymes in rat liver. II. Carbohydrate repression of dietary and hormonal induction of threonine dehydrase and ornithine δ-transaminase. J. Biol. Chem. 239: 4308-4313.

Potter, V.R. 1969. Recent trends in cancer biochemistry: The importance of studies on fetal tissue. Canadian Cancer Conference 8: 9-30.

Potter, V.R., M. Watanabe, H.C. Pitot, and H.P. Morris. 1969. Systematic oscillations in metabolic activity in rat liver and hepatomas. Survey of normal diploid and other hepatoma lines. Cancer Res. 29: 55-78.

Schimke, R.T. 1962. Adaptive characteristics of urea cycle enzymes in the rat. J. Biol. Chem. 237: 459-468.

Sheid, B. and J.S. Roth. 1965. Some effects of hormones and L-aspartate on the activity and distribution of aspartate aminotransferase activity in rat liver. Advances in Enzyme Regulation 3: 335-350.

Siperstein, M.D., V.M. Fagan, and H.P. Morris. 1966. Further studies on the deletion of the cholesterol feedback system in hepatomas. Cancer Res. 26: 7-11.

Taussky, H.H. 1954. A microcolorimetric determination of creatine in urine by the Jaffe reaction. J. Biol. Chem. 208: 853-861.

Weber, G., R.L. Singhal, and S.K. Srivastava. 1965. Regulation of RNA metabolism and amino acid level in hepatomas of different growth rate. Advances in Enzyme Regulation 3: 369-387.

Wu, C. 1964. Glutamine synthetase III. Factors controlling its activity in the developing rat. Arch. Biochem. Biophys. 106: 394-401.

Wu, C. 1965. Glutamine synthetase VI. Mechanism of the dithiol-dependent inhibition by arsenite. Biochim. Biophys. Acta 96: 134-147.

Wu, C. 1967. "Minimal deviation" hepatomas: A critical review of the terminology, including a commentary on the correlation of enzyme activity with growth rate of hepatomas. J. Natl. Cancer Inst. 39: 1149-1154.

Wu, C. and J.M. Bauer. 1960a. A study of free amino acids and of glutamine synthesis in tumor-bearing rats. Cancer Res. 20: 848-857.

Wu, C. and J.M. Bauer. 1960b. Metabolism of guanidoacetic acid in tumor-bearing rats. Proc. Soc. Exper. Biol. Med. 103: 422-424.

Wu, C. and J.M. Bauer. 1962. Catabolism of xanthine and uracil in tumor-bearing rats. Cancer Res. 22: 1239-1245.

Wu, C., J.M. Bauer, and H.P. Morris. 1971. Responsiveness of two urea cycle enzymes in Morris hepatomas to metabolic modulations. Cancer Res. 31: in press.

Wu, C. and H.A. Homburger. 1969. Responsiveness of enzymes in liver to growth of Novikoff hepatoma. British J. Cancer 23: 204-209.

Wu, C. and H.P. Morris. 1970. Responsiveness of glutamine metabolizing enzymes in Morris hepatomas to metabolic modulations. Cancer Res. 30: in press.

Wu, C., E.H. Roberts, and J.M. Bauer. 1965. Enzymes related to glutamine metabolism in tumor-bearing rats. Cancer Res. 25: 677-684.

Yeung, D. and I.T. Oliver. 1968. Factors affecting the premature induction of phosphopyruvate carboxylase in neonatal rat liver. Biochem. J. 108: 325-331.

BIOCHEMICAL ASPECTS OF ACCLIMATION TO A COLD ENVIRONMENT

Robert E. Beyer

Laboratory of Chemical Biology, Department of Zoology

The University of Michigan, Ann Arbor, Michigan 48104

Recent ENACT Teach-Ins at The University of Michigan, and elsewhere, have made it frighteningly apparent that our atmosphere is so endangered as to make life, as we know it, impossible on our planet in the forseeable future. One of the insidious aspects of the pollution problem involves the possibility of extreme alterations of the earth's ambient temperature, a problem which appears difficult to analyze and extrapolate to future decades. A continued increase in the carbon dioxide concentration of the earth's atmosphere would cause a "greenhouse effect", i.e. infra red radiated from the surface of the earth would be absorbed increasingly by atmospheric carbon dioxide, resulting in a progressive warming of the atmosphere (Plass, 1959; Keeling, 1970). On the other hand, scattering of the suns radiation by an increased atmospheric dust content would cause a loss of heat and light reaching the surface of the earth, resulting in a cooling of our atmosphere (Wexler, 1952). Of particular concern in this context is the possibility that in the mid-1980's approximately 500 American-built SST's, not to mention the French-British Concorde and the Soviet SST, will be flying at about 1800 miles per hour at between 65,000 and 70,000 feet. An MIT-Sponsored "Study of Critical Environmental Problems" (The Global Environment, 1970) reports "that the SST's may cause water vapor in the atmosphere to increase by 10 percent globally, or by as much as 60 percent over the North Atlantic, where SST traffic is expected to be heaviest." In addition, the SST's may cause stratospheric smog as a result of engine discharge of soot, hydrocarbons, nitrogen oxides, and sulfate particles. The fine particles thus formed could act to absorb and scatter radiations from the sun. Some geophysicists predict the coming of a new ice age. Assuming a pessimistic attitude toward the ability of our society to bring about realistic changes in public attitudes and governmental action with regard to air pollution it would appear to be highly desirable for this and future generations to be armed with an understanding of the basic responses of living systems to significant and chronic alterations in the temperature of the environment.

Toward this end we have, for some time, been involved in a study of the biochemical responses of a homeotherm, the laboratory rat, to a chronic cold environment. Before dealing with the specific biochemical systems which we have studied, it is important to present an overview of the general responses which enable the rat to maintain thermal equilibrium with a cold environment. Very soon after being placed in the cold, increased firing of peripheral thermoreceptors results in increased muscle tone and shivering, a mechanism which results in an increased rate of heat production and a phase in the adaptation of the organism called the "shivering phase of thermogenesis." The biochemical means by which heat is released at an increased rate during shivering thermogenesis is well understood. The major exergonic reactions in living cells are catalyzed by systems of mitochondrially bound oxido-reduction enzymes known as the electron transfer chain which is the final common pathway into which essentially all other catabolic pathways feed. As indicated in Figure 1, the end-product of carbohydrate, lipid and protein metabolism, catabolic sequences catalyzed in the soluble phase of the cell (cytosol), is an activated two-carbon unit, acetyl-CoA*. Acetyl-CoA is a uniquely important metabolic intermediate since it is able to interact with oxaloacetic acid, a four-carbon compound, to form citric acid, and thus enter the citric acid cycle sequence of reactions, including decarboxylations and dehydrogenations, which yield oxaloacetic acid, two carbon dioxides, and more important for our discussion, several molecules of reduced NAD* (NADH), and a molecule of succinic acid. The latter two compounds serve as oxidizable substrates for the electron transfer chain, also located within the mitochondrion. During the molecular manipulations of carbohydrates, fats and amino acids into and through the citric acid cycle, ending in succinic acid and NADH, most of the energy contained in the original molecules has been retained. From a teleological view, it would appear that the entire rationale for the reactions leading to the formation of succinic acid and NADH is to provide substrates of high negative oxidation-reduction potential for the electron transfer chain.

---

*The following abbreviations are used: CoA, coenzyme A; NAD, NADH, oxidized and reduced nicotinamide adenine dinucleotide; CoQ, coenzyme Q or ubiquinone; ADP, ATP, adenosine di- and triphosphate; P_i, inorganic orthophosphate.

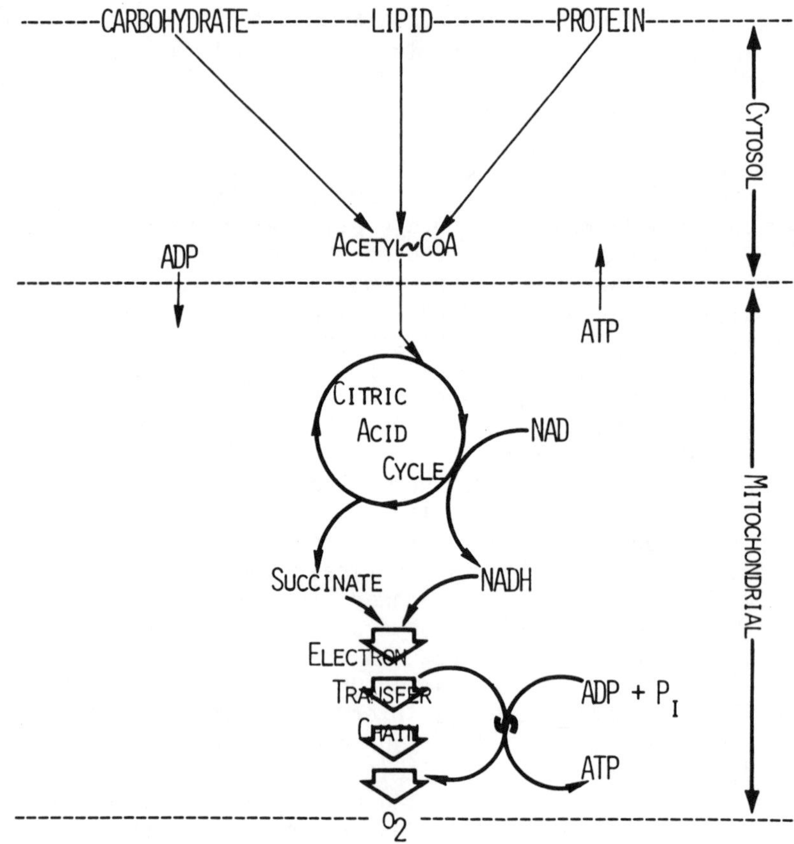

Figure 1. Cellular metabolic interrelations.

Thus, the major portion of the energetic content of food, after various molecular manipulations in the cell, is presented to the electron transfer chain in the form of molecules of rather negative oxido-reduction potential, ca. - 320 mv. Electrons are transferred, via the electron transfer chain components, to oxygen with an oxido-reduction potential of +820 mv.

Figure 2 contains a diagram showing a representation of the mitochondrial electron transfer chain. The two substrates of the electron transfer chain, NADH and succinic acid, are dehydrogenated by their respective dehydrogenases, both of which contain flavo-protein (FP) and non-heme iron (FeNH), arranged as individual structural complexes (complexes I and II respectively). A proton is released and an electron is transferred to a mobile component, coenzyme Q*, which acts to transfer electrons between the immobile

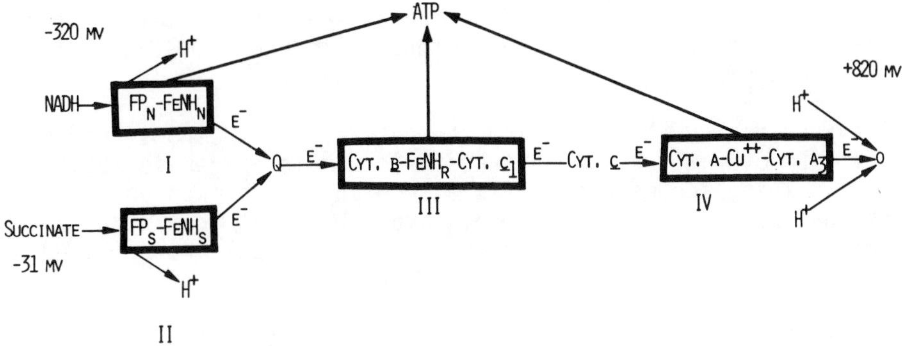

Figure 2. The mitochondrial electron transfer chain and oxidative phosphorylation.

Figure 3. Structure of coenzyme Q_{10}.

complexes I and II to complex III. The structure of CoQ (Figure 3), with its fifty-carbon isoprenoid side-chain, may allow mobility in the hydrophobic phospholipid environment of the mitochondrial membrane enabling the quinone nucleus to swing freely between the reducing complexes I and II and the oxidizing complex III. Complex III, comprised of cytochromes bc, and another nonheme iron, transfers electrons from reduced CoQ to oxidized cytochrome c. Due to its low molecular weight (12,500) and its ease of extraction from the mitochondrial membrane, cytochrome c is also considered a mobile component of the electron transfer chain, functioning as a reductant of the terminal complex IV, which contains cytochromes a, a_3, and copper. Complex IV, also called cytochrome oxidase, catalyzes the reduction of atomic oxygen. Two protons are added from the medium and the final product is water.

The free energy change during the passage of electrons from one mole of NADH over the electron transfer chain to oxygen is -52,000 calories. This becomes a significantly high figure when one considers that of the -686,000 cal/mole available from the oxidation of glucose to carbon dioxide and water, -624,000 cal are released at the level of the electron transfer chain.

The major role played by the electron transfer chain is to release energy from molecules derived from food. However, since living systems, being essentially isothermal, are not capable of utilizing heat to do useful work, a mechanism has evolved in animal systems which couples the release of energy during electron transfer with the endergonic synthesis of ATP*, the energy "currency" of the cell. This process, called oxidative phosphorylation, results in the formation of three moles of ATP/mole of NADH oxidized by oxygen. Since approximately 7000 cal are required to synthesize one mole of ATP from one mole each of ADP* and inorganic orthophosphate, and since a decline in free energy accompanying the oxidation of a mole of NADH by oxygen equals -52,000 cal, the efficiency of the overall oxidative phosphorylation process is 3 x 7000/52,000 = 40%.

As indicated in Figure 2, not all of the electron transfer chain complexes are capable of supporting the synthesis of ATP. The reason for this may be seen to be a thermodynamic limitation in complex II (Table 1) which has a free energy change of -4,600, well below the 7000 cal required for the synthesis of a mole of ATP. The remaining complexes I, III, and IV are energetically capable of supporting ATP synthesis, although complex III would appear to be at the limit of thermodynamic possibility.

TABLE 1

FREE ENERGY CHANGE IN THE ELECTRON TRANSFER CHAIN

Complex	Reaction	ΔG, cal/mole
I	NADH ----->CoQ	-19,000
II	Succinate ----->CoQ	- 4,600
III	$CoQH_2$ --------->cyt *c*+++	- 6,950
IV	cyt *c*++ ------->O_2	-25,500

One of the important control characteristics of the oxidative phosphorylation process is the inability of the electron transfer chain to transfer electrons in the absence of ADP, a phenomenon called respiratory control (Lardy and Wellman, 1952). Well-prepared isolated mitochondria oxidize substrate at near maximal rates in the presence of ADP (and Pi*). This may be seen in Figure 4 which contains a polarographic tracing of skeletal muscle mitochondria (Kuner and Beyer, 1970) during the oxidation of glutamate, a product of protein catabolism. When all of the added ADP has been phosphorylated to ATP, respiration is inhibited due to the coupling between the electron transfer and the phosphorylation reactions. A Respiratory Control Index may be calculated as the ratio of the rates of oxygen consumption during the two states. In the case of the particular experiment shown in Figure 4, the Respiratory Control Index was about 7.

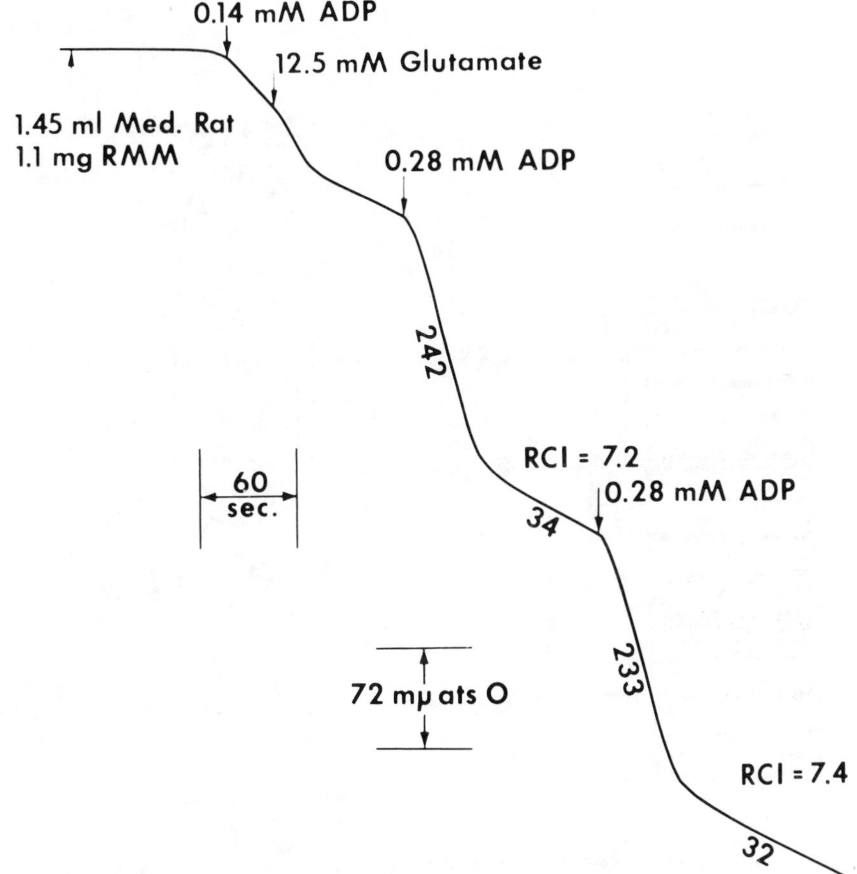

Figure 4. Respiratory control by ADP in isolated skeletal muscle mitochondria (From Kuner and Beyer, 1970).

The Respiratory Control Index is an important consideration in the process of heat liberation during the shivering phase of thermogenesis accompanying cold exposure. As mentioned above, temperature receptors inform the brain that a thermal gradient exists between the organism and the external environment, resulting in an increased rate of firing of motor fibers to the skeletal musculature. The immediate response is an alteration of the permeability characteristics of the sarco-plasmic system which releases calcium ions contained therein to the vicinity of the contractile structure (cf Figure 5). In the presence of calcium and ATP, the muscle contracts, utilizing the energy potential of ATP, resulting in the hydrolysis of ATP to yield ADP and Pi, yielding +7000

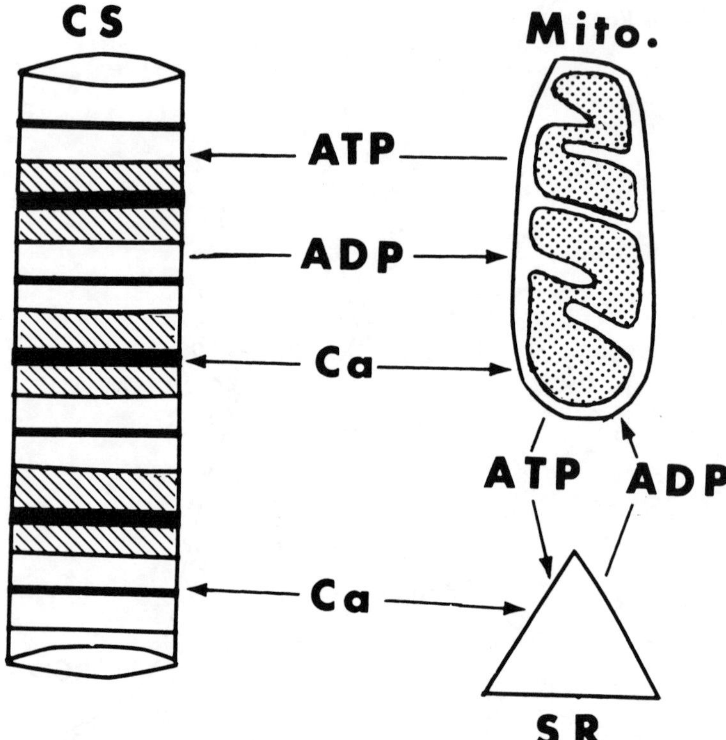

Figure 5. Diagramatic representation of ATP utilization and calcium movements in skeletal muscle. SR = sarcoplasmic reticulum; Mito. = mitochondrion; CS = contractile system.

cal/mole of ATP hydrolyzed. The muscle cannot relax, so as to be primed for another contraction cycle, in the presence of calcium. The concentration of calcium is lowered by the action of an endergonic, ATP-requiring calcium pump located on the sarco-plasmic reticulum which transports calcium across this membrane into the interior of the reticulum and away from the contractile mechanism. ADP and Pi, products of the two endergonic processes, serve as substrates for ATP synthesis, and at the same time, release the respiratory control restraints imposed upon the electron transfer chain by the absence of ADP. Thus, with shivering, ATP is utilized at an increased rate, resulting in an increased rate of oxidation of food stuffs and heat production. In the steady-state condition, essentially all of the energy released *via* electron transfer is converted to heat, either as the 60% released directly as heat, or *via* the hydrolysis of ATP during muscle contraction and calcium transport.

Before proceeding to a discussion of biochemical alterations which accompany chronic cold exposure, one might inquire as to whether mechanisms exist which aid the animal in its efforts to maintain ATP production during the initial, shivering phase. A well known response by mammals to acute stressful situations is the release of adrenal hormones. With respect to acute cold exposure it is interesting to note that liver mitochondria isolated from rats which have been injected with epinephrine are capable of maintaining their ATP synthetic capacity for extended periods as compared to controls (Lianides and Beyer, 1960a). Conversly, liver mitochondria prepared from adrenal-demedullated rats lose their phosphorylative capacity rapidly after isolation. It should be noted, however, that the phosphorylative efficiency and respiratory control of freshly prepared mitochondria from liver (Lianides and Beyer, 1960b) and skeletal muscle (Beyer, unpublished data) from rats during the shivering phase are normal. The data on mitochondrial stability from epinephrine treated and demedulated rats may reflect a compensatory mechanism which renders the mitochondrial system more resistant to extra mitochondrial factors during acute stress.

The initial phase of heat production in the cold is temporary, lasting approximately two to three weeks at 0-2°C, and may be viewed as an emergency tactic enabling the animal to survive a period during which more practical and permanent mechanisms are developed for increased theromogenesis. As indicated above, the second, or "non-shivering," phase of cold-induced thermogenesis is characterized by a continued high rate of heat production, and oxygen and food consumption, but a loss of high muscle tone and shivering (Hart et al., 1956; Heroux et al., 1956; Sellers et al., 1954). It has been shown that the non-shivering phase is qualitatively different from the shivering stage by its lack of sensitivity to the neuromuscular junction poison, curare (Cottle and Carlson, 1956).

CARL A. RUDISILL LIBRARY
LENOIR RHYNE COLLEGE

Since the shivering phase is dependent upon nervous activity, blocking nerve impulses from reaching the muscle results in a loss of shivering and, thus, a loss of heat from this process. Conversly, increased heat production during the non-shivering phase is not inhibited by curare, clearly indicating that a metabolic adjustment has occurred which releases heat independent of muscle contraction as an ATP utilizer.

Several proposals have been suggested to account for non-shivering thermogenesis at the biochemical level. One of the earliest proposals invoked partial physiological uncoupling of the phosphorylation of ADP from the electron transfer chain (Panagos and Beyer, 1958; Smith and Fairhurst, 1958a, b; Panagos et al., 1958). Under such conditions the electron transfer chain would be less constrained by the rate at which ADP was made available to the mitochondrion by the action of ATP-utilizing cellular systems. This, in turn, would enable heat to be released by the flow of electrons over the electron transfer chain at a greater rate and would place the control of the rate of heat release at the level of NADH production, i.e. substrate availability. Direct experimental tests of the "uncoupling" hypothesis, using isolated mitochondria from liver and skeletal muscle, have not been strongly supportive, however, although indirect tests indicate that a "loosening" of coupling may make some contribution to increased heat production (Smith and Fairhurst, 1958b; Panagos et al., 1958; Lianides and Beyer, 1960b), at least in liver systems.

Table 2 contains previously unpublished data on several parameters of mitochondrial energy metabolism in relation to cold acclimation using isolated rat skeletal muscle mitochondria as the test system. The first question which we asked was: Do rats synthesize more mitochondria during acclimation to cold in order to handle an increased substrate oxidation load? The yield of mitochondria per amount of protein used as starting material was

TABLE 2

YIELD OF SKELETAL MUSCLE MITOCHONDRIA DURING COLD-ACCLIMATION*

Days at 0-2°	mg mitochondrial protein/g tissue		
	cold	control	P
2	3.5	3.5	>.58
8	3.3	3.3	>.58
21	3.3	3.4	.15

*A total of 32 animals were used in this study and the data were analyzed statistically by the paired comparison method.

therefore assayed. The data in Table 2 clearly indicate that essentially identical amounts of mitochondria were isolated from skeletal muscle of control, acclimating (8 days) and cold-acclimated (21 days) rats. These data indicate that skeletal muscle does not contain greater amounts of mitochondrial protein as an adjustment to a new energetic situation.

The next question which we asked of these mitochondria was: Are such skeletal muscle mitochondria capable of an increased rate of electron transfer in either the controlled (minus ADP) or the released (plus ADP) state? Measurements were made as indicated in Figure 4, using two types of substrate and the data are recorded in Table 3. Statistically significant differences were noted in three of four of the cold-acclimated (21 days) tests, and in each case the rate of substrate oxidation by the mitochondrial preparations from cold-acclimated rats exceeded the rates of the control preparations. It is important to point out, however, that although the differences mentioned above were statistically significant, the quantitative differences were small and difficult to consider seriously as a basis for the 100% differences in heat output and oxygen consumption seen in intact animals. Such differences may be viewed as contributory to, but not as a major basis for, the difference in heat production between control and cold-acclimated animals.

TABLE 3

STATES 3 AND 4 RESPIRATION CATALYZED BY SKELETAL MUSCLE MITOCHONDRIA FROM CONTROL AND COLD-EXPOSED RATS*

Days at 0-2°	State 3					
	Pyruvate-Malate			Glutamate		
	Cold	Control	P	Cold	Control	P
2	277	277	>.58	255	275	.03
8	307	320	.36	304	316	.36
21	319	286	.05	340	319	.26
	State 4					
	Pyruvate-Malate			Glutamate		
	Cold	Control	P	Cold	Control	P
2	45.8	47.8	.12	41.0	42.1	.41
8	53.8	55.1	.36	47.2	45.1	.08
21	52.4	49.2	.05	44.4	41.6	<.01

*All rates are in terms of natoms oxygen consumed x min^{-1} x mg^{-1} protein and were analyzed by the paired comparison method.

The next question asked of these data concerned the phosphorylation efficiency (ADP/O ratio) and respiratory control of these same preparations: Do skeletal muscle mitochondria from cold-exposed rats exhibit a loosened respiratory control and does their efficiency of oxidative phosphorylation decrease, both consistent with a mechanism for increased thermogenesis in the cold? An analysis of the respiratory control ratios with two substrates, pyruvate-malate and glutamate, and three periods of cold-exposure, indicates that no significant differences between control and cold-exposed preparations were evident (Table 4). A slight, but significant increase in the efficiency of oxidative phosphorylation with glutamate as substrate was observed after eight days exposure to cold. This is in a direction opposite to that which would contribute to an increase in heat release and is not considered to be quantitatively important. From the data derived from the 32 animals used in the experiments reported in Tables 2 - 4 it may be concluded that skeletal muscle mitochondria isolated from cold-exposed and cold-acclimated rats do not display alterations in energy metabolism which might be interpreted as consistent with a mechanism for increased substrate utilization and heat production.

A smaller number of experiments have been performed on such preparations using NADH and alpha-glycerophosphate as substrates and no indications of differences between control and cold animals were observed.

TABLE 4

ADP/O AND RESPIRATORY CONTROL RATIOS OF SKELETAL MUSCLE MITOCHONDRIA FROM COLD-EXPOSED RATS*

Days at 0-2°	Respiratory control ratio (state 3/state 4)					
	Pyruvate-Malate			Glutamate		
	cold	control	P	cold	control	P
2	6.1	5.9	.19	6.3	6.7	.28
8	6.0	5.9	>.57	6.6	7.2	.12
21	6.1	5.8	.16	7.7	7.7	<.58
	ADP/O ratio					
	Pyruvate-Malate			Glutamate		
	cold	control	P	cold	control	P
2	2.33	2.35	.09	2.20	2.20	>.58
8	2.55	2.53	>.57	2.51	2.43	.006
21	2.62	2.48	.08	2.54	2.44	.23

*Data were analyzed for statistical significance by the paired comparison method.

In view of the demonstrations (Depocas, 1960a,b) that cold-acclimated rats respond preferentially to norepinepherine by increasing their heat production over and above their cold-acclimated rate, we have studied energetic parameters in skeletal muscle mitochondria isolated from rats after norepinepherine injection. A dosage of 0.4 mg of norepinepherine bitartrate per kg was used so as to induce approximately a doubling of heat production in the cold-acclimated rat. Control cold-acclimated rats were injected with an equivalent volume of 0.9% NaCl and after 30 minutes the rats were sacrificed and skeletal muscle mitochondria isolated. From the data in Table 5 it may be concluded that no consistent increase in any of the parameters of energy metabolism studied had been induced by norepinepherine. We are forced to conclude, for the time being, that if such a control which is reflected in a 100% increase in oxygen consumption in the cold animal exists *in vivo* at the mitochondrial level, it is either washed away during isolation of the mitochondrial preparation, or is dependent upon cellular integrity or extra mitochondrial factors for expression. It has been possible, however, to demonstrate (Panagos and Beyer, 1958) alterations of mitochondrial energy metabolism associated with cold-acclimation in isolated mitochondrial systems from liver. Such mitochondria exhibit lowered efficiency of oxidative phosphorylation (P/O ratios) with succinate as substrate beginning at about two weeks of cold exposure and continuing for at least three to four months in the cold (Table 6). These data indicate that a physical alteration has occurred in the mitochondrial membranes, possibly leading to an increased susceptibility of activated intermediate states to hydrolysis by water. An alternative explanation, based on

TABLE 5

EFFECT OF NOREPINEPHRINE ON ENERGY METABOLISM OF SKELETAL MUSCLE MITOCHONDRIA FROM COLD-ACCLIMATED RATS

	Cold control			Cold NE*-injected		
Days at 0-2°:	31	37	49	31	37	49
Resp. control:						
Pyruvate	5.0	6.0	7.0	5.3	5.4	6.2
Glutamate	6.4	6.5	8.4	6.2	6.8	8.0
ADP/O ratio:						
Pyruvate	-	2.4	2.2	-	2.4	2.2
Glutamate	-	2.4	2.2	-	2.4	2.0
State 3 resp:						
Pyruvate	382	344	552	428	346	480
Glutamate	389	340	554	414	368	526
State 4 resp:						
Pyruvate	77	57	79	81	64	78
Glutamate	61	52	66	67	55	66

*Norepinephrine

TABLE 6

OXIDATIVE PHOSPHORYLATION IN LIVER MITOCHONDRIA FROM COLD-ACCLIMATED RATS*

Days at 0-2°	P/O ratio	P
0 1	1.79 1.82	>0.58
0 14-16	1.71 1.26	0.013
0 90-120	1.71 1.49	0.011

*Data taken from Lianides and Beyer, 1960b.

TABLE 7

RESPIRATORY CONTROL BY LIVER MITOCHONDRIA FROM COLD-EXPOSED RATS WITH GLUTAMATE AS SUBSTRATE*

Days at 0-2°	Resp. control	P
0 1	2.94 3.03	0.55
0 13-16	2.91 2.29	0.06
0 27-32	2.97 2.74	<0.01
0 90-180	2.62 2.20	0.01

*Data from Lianides and Beyer, 1960b. These experiments were performed with a manometric apparatus.

the chemiosmotic hypothesis (Mitchell, 1968) would involve an alteration of the inner mitochondrial membrane resulting in an increased permeability to protons and a partial breakdown of the proton gradient generated by the electron transfer process. Regardless of the conceptual basis, such leaks or partial breakdown of the "activated state" would be expected to be expressed in a lowered respiratory control by the mitochondrial system. Table 7 contains data on the respiratory control of liver mitochondria from cold-exposed rats. The control drops significantly at two weeks of cold-exposure and remains lowered for at least six months. Although the loss of the control is not quantitatively great, it is significant and could form a contributing basis for increased heat production during cold-acclimation.

Additional evidence exists which indicates that liver mitochondria from cold-acclimated rats differ physically from those maintained at a more temperate environment. For example, such mitochondria swell more rapidly than do controls (Lianides and Beyer, 1960b), indicating that their ability to maintain their internal osmotic balance is impaired. In addition, their ability in vitro to maintain the synthesis of ATP is seriously impaired. Table 8 contains data (Lianides and Beyer, 1960b) which show that liver mitochondria isolated from rats exposed to 0-2° for periods of either two weeks or four to five weeks, when incubated in the absence of substrate, lose their ability to couple energy release via the oxidation of succinate to the synthesis of ATP considerably more rapidly than controls.

Liver mitochondria isolated from cold-acclimated rats are also more susceptable than control preparations to agents which uncouple oxidative phosphorylation. Figure 6 contains the results of experiments where the P/O ratio of liver mitochondria from control and one month cold-exposed rats were titrated against calcium ions. An increased susceptability to this uncoupler is evident and may indicate, again, a physical difference in membrane constituents or confirmation between mitochondria from control and cold-acclimated rats. It should also be mentioned that the difference in sensitivity

TABLE 8

LABILITY OF ATP SYNTHESIS IN LIVER MITOCHONDRIA FROM COLD-EXPOSED RATS*

Days at 0-2°	Aging index^{P}**	P
0	16.8	>0.57
1	17.8	
0	13.3	<0.01
14-16	6.3	
0	14.0	<0.01
28-35	8.3	

*Data from Lianides and Beyer, 1960b

**Defined as the time, in minutes, required to depress the phosphorylative capacity to a value halfway between the maximal and minimal rate of phosphorylation.

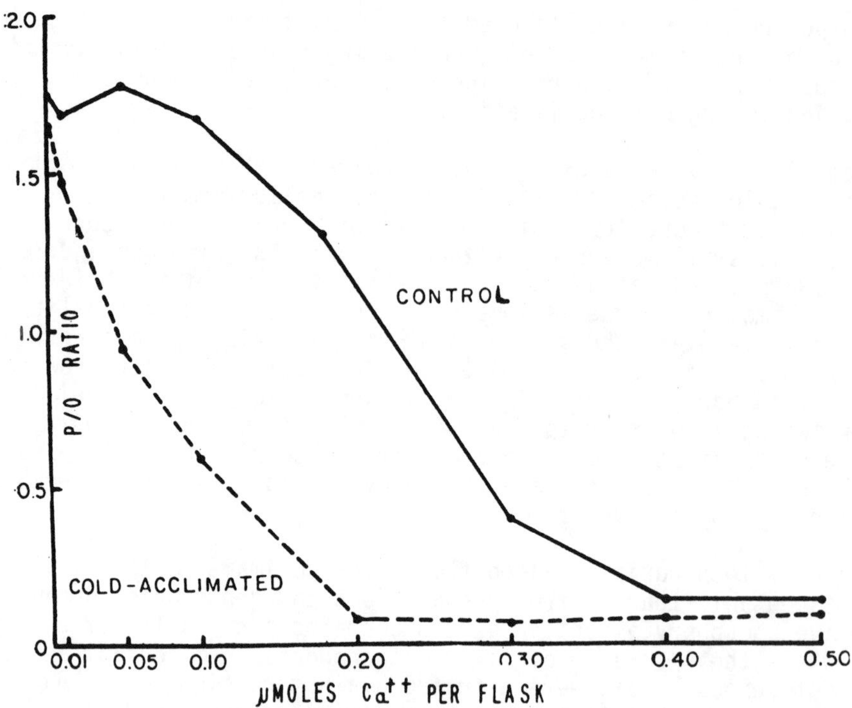

Figure 6. Effect of calcium on oxidative phosphorylation catalyzed by liver mitochondria from control rats and rats exposed to 0-2° for one month (From Lianides and Beyer, 1960b).

to calcium ions illustrated in Figure 6 was statistically highly significant as were results of experiments on rats maintained at 0-2° for two weeks and up to six months (Lianides and Beyer, 1960b).

At this point it might be well to examine the role of the thyroid in acclimation to cold. Rats in the cold appear to have increased thyroid activity (Cottle and Carlson, 1956; Woods and Carlson, 1956), and since the thyroid gland has been considered for many years as a major regulator of energy metabolism, it would not be unexpected to find thyroid involvement in cold acclimated animals. In addition, it has been suggested by many (Lardy and Feldott, 1951; Beyer et al., 1956; Ernster et al., 1959; Hoch, 1962) that the effect of thyroid hormones is based on its effect on mitochondrial systems. The presence of the thyroid gland has been shown to be required for the development of cold-acclimation as well as long-term survival in the cold (Leblond and Gross, 1943; Sellers et al., 1951). It was thus of interest to study the behavior of mitochondria from cold-acclimated rats which had been deprived of their thyroid hormone. Functional thyroidectomy was achieved by the injection of radioiodine into either control animals (25°) or rats maintained at 0-2° for 4-5 weeks. The injected animals were then maintained at the same environmental temperature for an additional three weeks prior to sacrifice and study of their liver mitochondria. Liver mitochondria from thyroidectomized rats were considerably more stable to the effects of incubation on loss of phosphorylative capacity than were controls when both were maintained at 25° (Table 9). As reported above, liver mitochondria from cold-acclimated rats lost their phosphorylative capacity more readily than room temperature controls. The interesting aspect of the experiment appears to be that mitochondria from cold-acclimated rats increase their stability following thyroidectomy, but to the level of room temperature controls instead of to room temperature, thyroidectomized preparations, indicating a control in the cold-acclimated rat in addition to the thyroid state. This phenomenon is, as yet, without an experimentally-based explanation.

In addition, and perhaps in conjunction with, a loosening of coupling between the phosphorylative and electron transfer functions of mitochondria, it has been suggested (Potter, 1958) that during cold-acclimation, electrons supplied by substrates involving extramitochondrial oxidations follow an alternate, poorly phosphorylating and poorly controlled, route, namely to the NADPH-cytochrome _c_ reductase pathway located at the level of the endoplasmic reticulum. Potter (1958) suggested that electrons from such extramitochondrial pathways might then enter the mitochondrial electron transfer chain at the level of mitochondrial cytochrome _c_ so as to communicate with oxygen. Since it is generally observed that the phosphorylation site associated with complex IV is "loosely coupled," this could conceivably provide for increased calorigenesis. In addition, some evidence exists which indicates that it is the first phosphorylation site, associated with complex I, which exerts the greatest degree of control over the flow of electrons through the electron transfer chain. Any pathway which could circumvent the first phosphorylation

TABLE 9

LABILITY OF PHOSPHORYLATIVE CAPACITY OF LIVER MITOCHONDRIA FROM THYROIDECTOMIZED COLD-ACCLIMATED RATS*

Environmental temperature	Aging Index^{P} Intact	Aging Index^{P} Athyroid
25°	14.2	>21.4
0-2°	8.3	13.6

*Data from Lianides and Beyer, 1960c

site could conceivably allow for increased rates of electron flow over the electron transfer chain relatively unencumbered by phosphorylative control. An early approach (Beyer, 1963) to the study of such alternate pathways involved relatively intact tissue slices of muscle and liver utilizing glucose as the substrate and inhibited by amytal. Amytal blocks the flow of electrons from NADH through complex I of the electron transfer chain (Ernster, 1955). The question asked was: Do tissues from cold-acclimated rats have an increased capacity to transfer reducing equivalents to oxygen using alternate routes to the highly controlled complex I? Table 10 contains data which indicate that the answer to the above question is "yes". A much larger percentage of the oxygen consumption is capable of bypassing the first phosphorylation site in the cold-acclimated tissues than in the control tissues. Unfortunately, the type of experiment shown in Table 10 does not provide any insight into the detailed alternate pathway which the cold-acclimated animal may possibly be using to overcome the tight metabolic control associated with the first phosphorylation site of the electron transfer chain; it merely suggests that such a possibility exists.

The question as to the precise pathway, or pathways, which might be regulated during cold-acclimation remains. An interesting possibility which deserved serious consideration was the DT diaphorase studied by Conover and Ernster (1962), formerly called vitamin K reductase (Märki and Martius, 1960). DT diaphorase is so named because it is a flavoprotein which catalyzes the transfer of electrons from either NADH or NADPH (formally DPNH and TPNH) to an electron acceptor in the electron transfer chain as depicted in Figure 6. Note that DT diaphorase is capable of transferring electrons from NADH or NADPH, *via* vitamin K3, to an electron acceptor of the electron transfer chain located on the oxygen side of the first phosphorylation site of Complex I. Although the precise

TABLE 10

EFFECT OF AMYTAL AN OXYGEN CONSUMPTION OF TISSUES OF COLD-ACCLIMATED (3 MONTHS) RATS*

Tissue	Oxygen consumption**, μat/100 mg		% amytal free
	Without amytal	With amytal	
Diaphragm			
Control	12.43	2.18	17.5
Cold	17.81	5.72	32.1
Liver slice			
Control	15.64	3.44	22.0
Cold	16.96	7.14	43.7

*Data from Beyer, 1963.
**Glucose as substrate

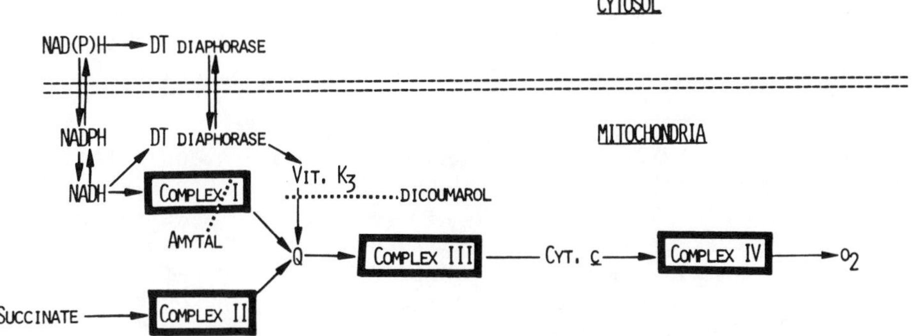

Figure 7. Possible pathways of electron transfer between DT diaphorase and the mitochondrial electron transfer chain.

TABLE 11

DT DIAPHORASE ACTIVITY OF LIVER MITOCHONDRIA FROM COLD-ACCLIMATED RATS*

Additions	μatoms O/200 mg eq mito/20 min	
	Control	Cold
None	6.21	6.44
Dicoumarol	6.09	5.27
Vitamin K_3	6.37	6.94
Amytal	0.39	0.91
Amytal + Vit K_3	6.16	8.60
Amytal + K_3 + Dicoumarol	0.99	1.15

*Data from Beyer, 1963.

electron acceptor of DT diaphorase has not been identified we have suggested coenzyme Q for this function in Figure 7 for illustrative purposes. DT diaphorase activity has been assayed in control and cold-acclimated rats (Beyer, 1963), using endogenous NADH generated by endogenous glutamic dehydrogenase with glutamate as substrate. The data (Table 11) support the contention that DT diaphorase activity is elevated in the cold-acclimated rats. In the presence of amytal and vitamin K3, DT diaphorase activity is greater in liver mitochondria from cold-acclimated rats by a factor of 1.4, indicating that this pathway may be contributing to increased thermogenesis in the cold-acclimated rat. As indicated in Figure 7, an extramitochondrial DT diaphorase also exists in the cytosol which may possibly function to shuttle reducing equivalents from extramitochondrially generated NADH and NADPH to the mitochondrial electron transfer chain, again bypassing the first phosphorylation site. The possibility that extramitochondrial DT diaphorase activity is also elevated during cold-acclimation has not been investigated, but deserves consideration.

Additional pathways for the transfer of reducing equivalents from extramitochondrially produced NADH to the mitochondrial electron transfer chain are possible. For example, we have investigated (Beyer, 1963) the activity of NADH-cytochrome c reductase activity of liver microsomes from control and cold-acclimated rats (Table 12). The activity of this microsomal electron transferring pathway was approximately doubled after thirty days of exposure to 0-2° for thirty days. Unfortunately, we have little information about whether a connection exists between the microsomal NADH-cytochrome c reductase and the mitochondrial electron transfer chain. Hopefully, future research will lead to the identification of cytosol factors capable of functionally linking such electron transfer pathways located in separate cellular compartments.

In a consideration of possible alternate pathways of electron transfer from the cytosol to the mitochondrial electron transfer chain, it is important to consider possible mitochondrial acceptors for reducing equivalents generated outside of the mitochondrion. Because of the unique arrangements between the immobile complexes of the electron transfer chain, their ability to serve as communication between complexes, and because of their mobility within a mitochondrion, the obvious candidates for such a function would be the mobile electron transfer components, NADH, coenzyme Q, and cytochrome c. The first candidate, NADH, would not appear to be a likely suspect since electrons from endogenous NADH must pass through the first phosphorylation site of complex I which, as I have mentioned above, is a highly controlled energy coupling site. Since the reoxidation of reduced coenzyme Q and cytochrome c would bypass the first phosphorylation site, these two components are more likely to serve in such a role. It would also be logical to suppose that if

TABLE 12

MICROSOMAL NADH-CYTOCHROME c REDUCTASE ACTIVITY FROM COLD-EXPOSED RATS*

Animal	mμmoles cyt. c reduced/min/mg N
Control	11.33
Cold	20.73
% change	+45.3

*Data from Beyer, 1963

coenzyme Q and cytochrome c were to serve such a function, the mitochondrial membranes interposed between these immobile electron transfer chain components and the cytosol might become altered in their permeability so as to result in greater accessibility between the chain and cytosol reducing factors and, in addition, that the mobile electron transfer chain components might be induced to higher steady-state concentrations to meet additional demands. Little direct information is available on the former possibility except for studies on the more rapid swelling of mitochondria from cold-acclimated mitochondria (Lianades and Beyer, 1960b), but data do exist on the latter point. For example, Klain (1963) has reported that cytochrome c levels are increased in heart, kidney, liver, lung, skeletal muscle, and spleen from cold-acclimated rats. After twenty days of exposure to 5°, cytochrome c levels were doubled in skeletal muscle, and tripled in liver, indicating that such a metabolic adaptation to cold was widespread and quantitatively significant. Depocas (1966) has verified the increase in skeletal muscle cytochrome c concentrations during cold-acclimation, but has reported smaller quantitative differences. Depocas (1966) has also reported that the turnover of cytochrome c is not affected by temperature. The other mobile component in the interior of the electron transfer chain, coenzyme Q, has also been shown to be increased in concentration in a number of tissues as a result of cold exposure (Beyer et al, 1962a), as well as by a number of other conditions which induce hyper-metabolism (Beyer et al, 1962b). This increase in coenzyme Q levels in cold-acclimated rats has been shown recently to be due to a decrease in the rate of catabolism during the entire period of cold exposure and an increase in the synthesis of coenzyme Q during the acclimating period (Aithal et al, 1968). It is interesting, and consistent with the idea expressed above on the necessity to by-pass the first phosphorylation site, that during acclimation to cold, NADH, the remaining mobile component of the electron transfer chain, is not increased in concentration (Hannon and Rosenthal, 1963; Smith and Fairhurst, 1958b).

In addition to the establishment of alternate pathways of electron transfer, and increases in normal thermogenic pathways, which are responsible for increased heat production during acclimation to cold, the possible involvement in thermogenesis of increased turnover of ATP by cellular endergonic reactions should be considered. For example, it is possible that the membranes of the sarcoplasmic reticulum system of muscle are so altered as a result of increased activity during shivering that they become "leaky" to calcium ions, requiring additional ATP to enable the calcium pump to lower the calcium concentration sufficiently so as to allow relaxation of muscle contractile units. The increased utilization of ATP would require the oxidative phosphorylation system to function at increased rates to supply the elevated requirement for ATP and would result in a corresponding increase in electron transfer and heat production. This possibility has not received rigorous testing.

Another possible mechanism for increased heat production during acclimation to cold could involve an increase in the highly endergonic turnover of proteins in the cell. An increase in the activity of intracellular proteolytic enzymes could release feedback inhibition and trigger protein synthetic activity, a process which requires both ATP and GTP at various points in the overall pathway. Since GTP is formed from GDP and ATP _via_ the action of a nucleotide diphosphokinase, the utilization of GTP, as well as ATP, would result in an increased level of electron transfer activity at the mitochondrial level. That an increase in protein turnover is, in fact, occurring in cold-acclimated animals is indicated by the increase in urinary nitrogen excretion of rats maintained at 2° for 11 weeks (Mefferd, 1960). In addition, Trapani (1960) has reported a 50% increase in protein turnover during cold exposure. More recently, the protein turnover rates of cold-acclimated rats has been reported to be increased (Yousef and Chaffee, 1970). Although the role of protein turnover in cold-induced thermogenesis has not been conclusively shown, the results are indicative and this possibility deserves further attention.

A discussion of thermogenesis during cold-acclimation would not be complete without a discussion of the role of brown fat, a subject which has been reviewed recently (Smith and Horowitz, 1969) and has received considerable attention (Nine papers, 1970). The interscapular region contains brown fat which is richly vascularized and contains an abundance of mitochondria. Upon exposure to cold the amount of brown fat tissue increases, as does its blood supply and heat production. This is significant because the brown fat tissue is so located that the heat it releases is utilized, _via_ the circulatory system, to warm such important functional areas as the heart and cervical spinal cord which are so important in the maintenance of temperature homeostasis. Brown adipose tissue responds to norepinephrine, whose secretion is increased during cold-exposure, by increasing its heat output. This appears to be accomplished by a stimulation of triglyceride hydrolysis by norepinepherine, resulting in the release of fatty acids from their storage form. The presence of free fatty acids appears to have a dual effect on mitochondria: 1) to loosen the coupling between phosphorylation and electron transfer and 2) to supply substrate to the beta-oxidation pathway of fatty acid metabolism which, in turn, supplies reduced substrate to the electron chain. The result, then, is that the respiratory control limitation of electron transfer is lifted and substrate for heat production from the transfer of reducing equivalents over the electron transfer chain is supplied. Although the total amount of heat supplied from brown adipose tissue metabolism is not quantitatively large in terms of the total increase in heat production during cold acclimation, it is qualitatively significant in terms of the important regulatory organs to which it supplies heat.

The identification of a single process responsible for increased heat production during cold-acclimation has proven to be an elusive goal. The possibility, that an elevation in the activity of a number of metabolic endergonic processes may together contribute to the cold-induced increase in heat production, should be seriously considered. It should be kept in mind, however, that it is at the level of the mitochondrial electron transfer chain that there exists an opportunity for the cell to release directly free energy bound in the configuration of oxidizable biological molecules. Essentially all biological thermogenic processes will have as their basis an increase in the rate of electron transfer at the mitochondrial level.

ACKNOWLEDGMENTS

Previously unpublished work reported herein was supported by grant AM 10056 from the National Institute of Arthritis and Metabolic Diseases of the National Institute of Health.

REFERENCES

Aithal, H.N., V.C. Joshi, and T. Ramasarma. 1968. Effect of cold exposure on the metabolism of ubiquinone in the rat. Biochim. Biophys. Acta 162:66-72.

Beyer, R.E. 1963. Regulation of energy metabolism during acclimation of laboratory rats to a cold environment. Fed. Proc. 22: 874-880.

Beyer, R.E., H. Löw, and L. Ernster. 1956. Effect of thyroxine on mitochondrial stability. Acta Chem. Scand. 10:1039-1041.

Beyer, R.E., W.M. Noble, and T.J. Hirschfeld. 1962a. Coenzyme Q (ubiquinone) levels of tissues of rats during acclimation to cold. Canad. J. Biochem. Physiol. 40:511-518.

Beyer, R.E., W.M. Noble, and T.J. Hirschfeld. 1962b. Alterations of rat-tissue coenzyme Q (ubiquinone) levels by various treatments. Biochim. Biophys. Acta 57:376-379.

Conover, T.E. and L. Ernster. 1962. DT diaphorase. II. Relation to respiratory chain of intact mitochondria. Biochim. Biophys. Acta 58:189-200.

Cottle, W.H. and L.D. Carlson. 1956a. Regulation of heat production in cold-adapted rats. Proc. Soc. Exptl. Biol. Med. 92: 845-849.

Cottle, M. and L.D. Carlson. 1956b. Turnover of thyroid hormone in cold-exposed rats determined by radioactive iodine studies. Endocrinol. 59:1-11.

Depocas, F. 1960a. The calorigenic response of cold-acclimated white rats to infused noradrenaline. Canad. J. Biochem. Physiol. 38:107-114.

Depocas, F. 1960b. Calorigenesis from various organ systems in the whole animal. Fed. Proc. 19, Suppl. 5:19-24.

Depocas, F. 1966. Concentration and turnover of cytochrome c in skeletal muscles of warm- and cold-acclimated rats. Canad. J. Physiol. Pharm. 44:875-880.

Ernster, L., D. Ikkos, and R. Luft. 1959. Enzymic activities of human skeletal muscle mitochondria: A tool in clinical metabolic research. Nature 184:1851-1854.

Ernster, L., O. Jalling, H. Löw, and O. Lindberg. 1955. Alternative pathways of mitochondrial DPNH oxidation, studies with amytal. Exptl. Cell Research, Suppl 3:124-132.

Hannan, J.P. and A. Rosenthal. 1963. Effects of cold acclimatization on liver di- and triphosphopyridine nucleotide. Am. J. Physiol. 204:515-516.

Hart, J.S., O. Heroux, and F. Depocas. 1956. Cold acclimation and the electromyogram of unanesthetized rats. J. Appl. Physiol. 9:404-408.

Heroux, O., J.S. Hart, and F. Depocas. 1956. Metabolism and muscle activity of anesthetized warm and cold acclimated rats on exposure to cold. J. Appl. Physiol. 9:399-403.

Hoch, F. 1962. Thyrotoxicosis as a disease of mitochondria. New England J. Med. 266:446-454.

Keeling, C.D. 1970. Is carbon dioxide from fossil fuel changing man's environment? Proc. Am. Philos. Soc. 114:10-14.

Klain, G.J. 1963. Alterations of rat-tissue cytochrome c levels by a chronic cold exposure. Biochim. Biophys. Acta 74:778-780.

Kuner, J.M. and R.E. Beyer. 1970. An ultrastructural study of isolated rat skeletal muscle mitochondria in various metabolic states. J. Membrane Biol. 2:71-84.

Lardy, H.A. and G. Feldott. 1951. Metabolic effects of thyroxine *in vitro*. Ann. N.Y. Acad. Sci. 54:636-648.

Lardy, H.A. and H. Wellman. 1952. Oxidative phosphorylations: Role of inorganic phosphate and acceptor systems in control of metabolic rates. J. Biol. Chem. 195:215-224.

Leblond, C.P. and J. Gross. 1943. Effect of thyroidectomy on resistance to low environmental temperature. Endocrinol. 33: 155-160.

Lianides, S.P. and R.E. Beyer. 1960a. Oxidative phosphorylation in liver mitochondria from adrenal-demedullated and epinephrine-treated rats. Biochim. Biophys. Acta 44:356-357.

Lianides, S.P. and R.E. Beyer. 1960b. Oxidative phosphorylation in liver mitochondria from cold-exposed rats. Am. J. Physiol. 199:836-840.

Lianides, S.P. and R.E. Beyer. 1960c. Thyroid function and oxidative phosphorylation in cold-exposed rats. Nature 188: 1196-1197.

Märki. F. and C. Martius. 1960. Vitamin K-Reduktase, Darstellung und Eigenschaften. Biochem. Z. 333:111-135.

Mefferd, R.B., Jr. 1960. Comments on intermediary metabolism in relation to cold. Fed. Proc. 19, Suppl. 5:121-124.

Mitchell, P. 1968. Chemiosmotic coupling and energy transduction. Glynn Research Ltd., Bodmin, England.

Panagos, S. and R.E. Beyer. 1958. Oxidative phosphorylation and stability of liver mitochondria from rats exposed to low environmental temperatures. Fed. Proc. 17:121.

Panagos, S., R.E. Beyer, and E.J. Masoro. 1958. Oxidative phosphorylation in liver mitochondria prepared from cold-exposed rats. Biochim. Biophys. Acta 29:204-205.

Plass, G.N. 1959. Carbon dioxide and climate. Sci. Am. 201; No. 1:41-47.

Potter, V.R. 1958. Possible biochemical mechanisms underlying adaptation to cold. Fed. Proc. 17:1060-1063.

Sellers, E.A., J.W. Scott, and N. Thomas. 1954. Electrical activity of skeletal muscle of normal and acclimated rats on exposure to cold. Am. J. Physiol. 177:372-376.

Sellers, E.A., S.S. You, and N. Thomas. 1951. Acclimatization and survival of rats in the cold: Effects of clipping, of adrenalectomy and of thyroidectomy. Am. J. Physiol. 165:481-485.

Smith, R.E. and A.S. Fairhurst. 1958a. Cellular mechanisms of cold adaptation in the rat. Fed. Proc. 17:151.

Smith, R.E. and A.S. Fairhurst. 1958b. A mechanism of cellular thermogenesis in cold-adaptation. Proc. Natl. Acad. Sci.(U.S.) 44:705-711.

Smith, R.E. and B.A. Horwitz. 1969. Brown fat and thermogenesis. Physiol. Rev. 49:330-425.

The Global Environment: M.I.T. Study Looks for Danger Signs. 1970. Science 169, 660-662.

Trapani, I.L. 1960. Cold exposure and the immune response. Fed. Proc. 19, Suppl. 5:109-114.

Various authors. 1970. Nine papers published as a symposium on "Brown Adipose Tissue." Lipids 5, #1.

Wexler, H. 1952. Volcanoes and world climate. Sci. Am. 186, No. 4:74-80.

Woods, R. and L.D. Carlson. 1956. Throxine secretion in rats exposed to cold. Endocrinol. 59:323-330.

Yousef, M.K. and R.R.J. Chaffee. 1970. Studies on protein turnover rates in cold-acclimated rats. Proc. Soc. Exptl. Biol. Med. 133:801-804.

THE EFFECTS OF CHLORINATED HYDROCARBONS ON THE HEPATIC MONOOXYGENASE SYSTEM

David Kupfer

Lederle Laboratories, Division of American Cyanamid Company

Pearl River, New York 10965

Monooxygenases, also referred to as mixed function oxidases, are enzymes which in the presence of an electron donor and molecular oxygen catalyze the incorporation of oxygen into aliphatic and aromatic substrates. The stoichiometry of this reaction can be described as follows:

$$AH + O_2 + BH + H^+ \longrightarrow AOH + B + H_2O$$

where AH is the substrate and BH is the electron donor.

Among the various monooxygenases studied, particular emphasis has been placed on a monooxygenase system which resides in the tubular structure of the mammalian liver, the endoplasmic reticulum. In the course of homogenization of the liver and differential centrifugation, the endoplasmic reticulum is disrupted and the bulk of the NADPH - dependent monooxygenase activity is found in the microsomal fraction (100,000 x g precipitate). The microsomal monooxygenase activity is a system composed of several components which catalyze the electron transport and substrate oxidation. The currently envisioned scheme is as follows:

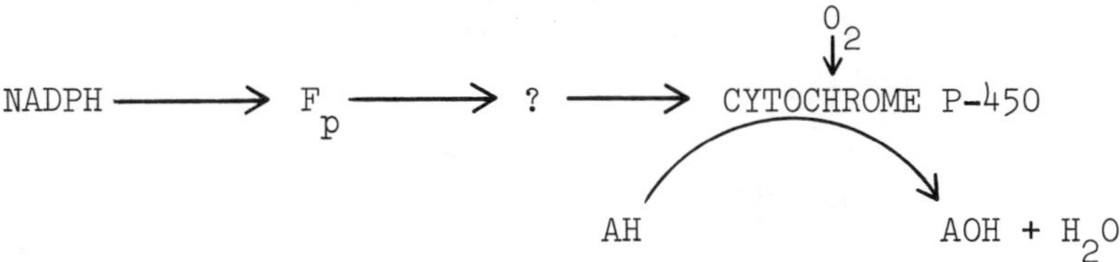

The oxidation of a substrate AH requires an electron flow from NADPH to a flavoprotein, via a possible unknown component, to cytochrome P-450 which binds both AH and O_2 and catalyzes the incorporation of oxygen into AH.

This monooxygenase enzyme system catalyzes the oxidative metabolism of a variety of lipid soluble substances, among them drugs, and has been often referred to as drug-metabolizing enzymes (Conney, 1967). Moreover, it was found that various factors such as sex and age of the animal and exposure to stress influence the activity of this enzyme system.* A most interesting feature of the monooxygenase system is its ability to be induced.** For instance, the administration of structurally unrelated compounds into animals can cause a several-fold increase in enzymatic activity. Among the most potent inducers are barbiturates, which cause an increase in the metabolism of subsequently administered structurally similar or dissimilar drugs; this phenomenon is often accompanied by a diminished biological activity of these drugs.

The concern over the possible toxic effects of chlorinated hydrocarbon insecticides and the observations that these substances tend to accumulate in adipose tissue of animals of various species, would have probably led to the eventual discovery that chlorinated hydrocarbons are potent inducers of the microsomal monooxygenase system. However, the accidental discovery of this phenomenon preceded the "logical" approach. Hart et al (1963) observed that spraying of animal rooms with chlordane shortened the duration of barbiturate-induced sleep in the rat. Furthermore, the administration of chlordane to rats increased the oxidative metabolism of this barbiturate by liver microsomes. In the same year, Hart and Fouts (1963) showed that DDT, similarly to chlordane, also stimulates the activity of the hepatic monooxygenase system. Subsequently, Ghazal et al (1964) demonstrated with a single injection of a chlorinated hydrocarbon insecticide that there is a correlation between the duration of the effects on shortening hexobarbital sleep and on stimulation of hexobarbital oxidation. Gerboth and Schwabe (1964) demonstrated the sensitivity of the monooxygenase system to DDT; as little as 1 mg DDT/kg body weight shortened hexobarbital sleep.

---

* Higher enzymatic activity was observed in male rats than female rats. Adult animals have higher enzymatic activity than immature animals.

** Though induction has not been proven, in this article this term will be used synonymously with an increase in enzyme activity.

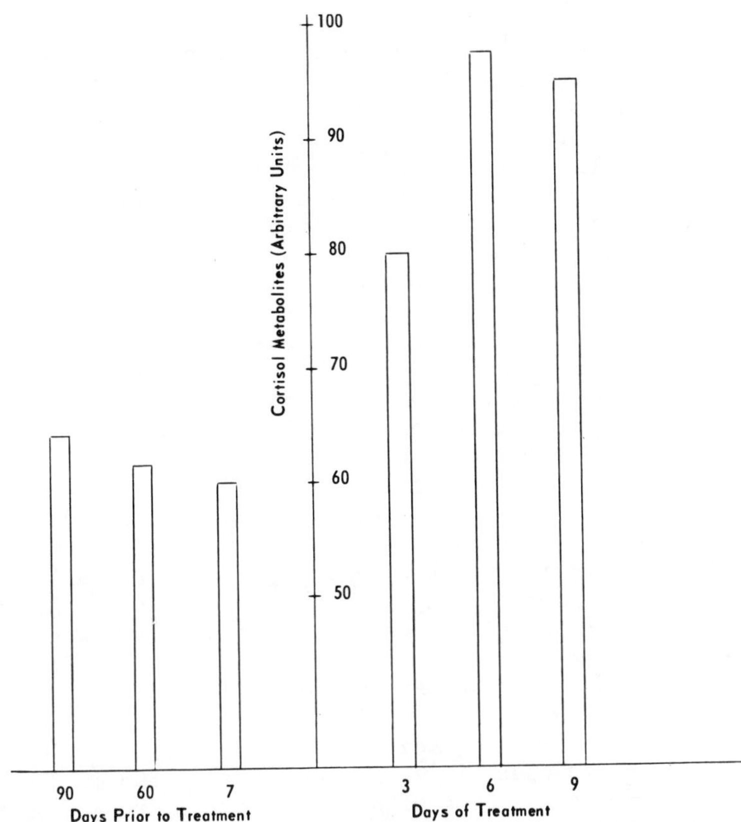

Fig. 1. The effect of oral administration of o,p'DDD (300 mg/kg for the first 5 days followed by 50 mg/kg on subsequent days) on urinary polar cortisol metabolites in the guinea pig.

Our interest in the effects of chlorinated hydrocarbons on the microsomal monooxygenase system evolved quite accidentally. Previous observations that o,p'DDD* [2,2-(<u>o</u>-chlorophenyl,<u>p</u>-chlorophenyl)-1,1-dichloroethane] is adrenotoxic in the dog (Cueto and Brown, 1958) and that o,p'DDD is useful in the treatment of adrenal overproduction of cortisol in man (Bergenstal et al., 1960), prompted us to investigate the effect of this compound on the adrenal production of cortisol in the guinea pig. To our surprise, o,p'DDD did not diminish cortisol formation, but increased the urinary excretion of polar (hydroxylated) cortisol metabolites which were derived from endogeneous cortisol and from administered ^{14}C-cortisol (Fig. 1) (Kupfer et al, 1964). Similar results were

---

* o,p'DDD is a contaminant present in preparations of technical grade DDT.

obtained when DDT was administered to guinea pigs (Balazs and Kupfer, 1966). These findings suggested that o,p'DDD and DDT stimulate cortisol hydroxylation, possibly by inducing the hepatic monooxygenase system. In fact, the activity of the hepatic monooxygenase system as measured in vitro by the rate of demethylation of p-chloro-N-methylaniline* was enhanced in guinea pigs treated with either o,p'DDD or DDT (Kupfer et al, 1969a). Further studies with o,p'DDD in rats showed that o,p'DDD, similarly to DDT, shortens hexobarbital-induced sleep (Table 1) and enhances the activity of the hepatic monooxygenase system (Table 2).

The rate determining step in the monooxygenase reactions is not known. Some evidence points to the rate of reduction of the cytochrome P-450 substrate complex as being the rate determining step (Gigon et al, 1969; Kupfer and Orrenius, 1970). Thus, phenobarbital produces an increase in both the concentration of cytochrome P-450 and in the rate of reduction of this cytochrome (Remmer and Schenkman, 1967; Gnosspelius et al, 1969/1970). Similarly, DDT was found to increase the concentration of microsomal cytochrome P-450 (Remmer and Schenkman, 1967). The mechanism of induction of the monooxygenase system is not understood. Studies with barbiturates and inhibitors of protein synthesis suggest that the stimulation of the monooxygenase system by barbiturates is probably due to enhanced protein formation (Conney, 1967). However, to my knowledge no such studies have been carried out with the chlorinated hydrocarbons. Nevertheless, though the evidence is incomplete, it appears that the mechanism of induction of the microsomal monooxygenase system with chlorinated hydrocarbons is similar to that obtained with barbiturates.

Recently, evidence has been brought forth that the hepatic monooxygenase system is probably involved in the oxidative metabolism of endogeneous fatty acids and steroids (Kuntzman et al, 1964; Kuntzman et al, 1965; Kupfer, 1968; Kupfer and Orrenius, 1970; Tephly and Mannering, 1968). The recognition that steroids may be the "natural" substrates for the hepatic monooxygenase system suggested that the alteration of the activity of this enzyme system by drugs or chlorinated hydrocarbons may have marked physiological effects in the animal.

Two groups of investigators independently approached this problem. The group at Burroughs Wellcome examined three biological parameters in the rat: (i) the duration of sleep induced by steroids,

---

* This demethylation is carried out by the monooxygenase system, the products formed are p-chloroaniline and formaldehyde (Kupfer and Bruggeman, 1966).

TABLE 1

EFFECT OF DDT AND O,P'DDD ON THE DURATION OF HEXOBARBITAL INDUCED SLEEP IN THE IMMATURE MALE AND FEMALE RAT

Treatment	Sleep (Min.)	
	Male	Female
Control	75	133
p,p'DDT	8	N.D.
o,p'DDD	19	25

p,p'DDT (200 mg/kg) and o,p'DDD (300 mg/kg) were given orally for 3 days. On the 4th day, hexobarbital (125 mg/kg) was injected intraperitoneally (i.p.). N.D. = not determined.

TABLE 2

THE EFFECT OF O,P'DDD ON HEXOBARBITAL METABOLISM BY 9,000 X G LIVER SUPERNATANT FROM IMMATURE AND ADULT MALE RATS

Animals		Hexobarbital Metabolism μmoles/g liver/30 min
Immature	Controls	0.75 ± 0.13
	Treated	3.54 ± 0.25
Adult		
	Controls	3.03 ± 0.23
	Treated	4.60 ± 0.20

Immature rates (60 g) were treated orally with 300 mg o,p'DDD/kg for 3 days. Adult rats (200 g) were treated with 300 mg o,p'DDD/kg for 9 days, controls received vehicle alone. Values represent mean ± standard error.

(ii) the increase in uterine weight (uterotropic effect) by estrogens and (iii) the androgen-induced weight increase of seminal vesicles. Their findings demonstrated that pre-treatment of rats with DDT, chlordane or phenobarbital entirely eliminated the sleep which can be produced by the administration of various steroids (cf Conney, 1967). An increase in the microsomal hydroxylation of these steroids was also observed. These investigators also demonstrated that pretreatment of rats with chlordane or phenobarbital diminished the uterotropic effect of estradiol in immature female rats and diminished the androgen induced increase in seminal vesicle weight in male rats. Such treatment with chlordane and phenobarbital also increased the <u>in vitro</u> microsomal hydroxylation of estrogens and androgens (Conney, 1967; Kuntzman, 1969; Levin et al 1969).

More recently an interesting observation was described in a preliminary communication by Fahim et al (1970) that a 10-day treatment with DDT (15 mg/kg) of <u>intact</u> and <u>castrate</u> mature female rats diminished the uterotropic activity and uterine alkaline phosphatase-inducing activity of administered estrogens.

Whereas other investigators studied primarily the effects of steroids on target organs, our efforts were concentrated on an effect by a given steroid on a specific liver enzyme. Cortisol and certain structurally related steroids are known to produce an increase in the activity of liver tyrosine transaminase (Lin and Knox, 1957; Sereni et al 1959; Kupfer, 1968). Our approach was directed to the search for a steroid which, by contrast to cortisol*, is metabolized primarily by the monooxygenase system, yielding a hydroxylated metabolite which is biologically inactive; i.e., does not induce tyrosine transaminase.

The steroid, triamcinolone acetonide (TrA) was found to satisfy these requirements. TrA is converted by rat liver microsomes almost exclusively into 6β-hydroxy-triamcinolone acetonide (6β-OH-TrA) (Fig. 2) which possesses little or no biological activity (Kupfer et al 1969b; Kupfer and Partridge, 1970). This transformation of TrA is catalyzed by a monooxygenase-type system as evidenced by the requirements for O_2 and NADPH and by the phenobarbital stimulation and SKF 525A inhibition of the formation of 6β-OH-TrA (Table 3). Furthermore, the treatment of rats with o,p'DDD was found also to enhance the transformation of TrA into 6β-OH-TrA (Table 4).

The question whether stimulation of the monooxygenase system

*Cortisol is metabolized primarily by ring A reduction; this metabolic pathway is not significantly affected by inducers of the monooxygenase system.

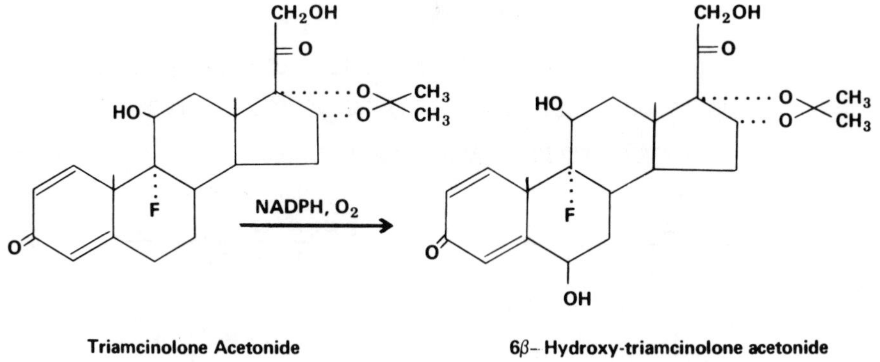

Triamcinolone Acetonide NADPH, O_2 6β-Hydroxy-triamcinolone acetonide

Fig. 2. Transformation of triamcinolone acetonide by hepatic monooxygenase system.

TABLE 3

CHARACTERISTICS OF RAT HEPATIC STEROIDAL 6β-HYDROXYLASE SYSTEM

	Treatment[a]	Incubation: Additions or Omissions	Gas Phase	6β-OH-TrA formed in 20 min. (nmoles)
1.	Control		Air	89
	Control	+SKF 525A (10^{-4}M)	Air	29
2.	Control		Air	71
	Control	-NADPH generating system	Air	24
3.	Phenobarbital		Air	130
	Phenobarbital		Air(-CO_2)	124
	Phenobarbital		N_2	32
	Phenobarbital	+SKF 525A	Air	79

[a]Male rats were injected i.p. with phenobarbital (37.5 mg/kg in 0.2 ml H_2O twice daily) for three days; controls were injected with H_2O.

Each incubation mixture contained 575 nmoles of ^{3}H-TrA. The incubation mixture and conditions were as previously described (Kupfer et al. 1969b).

TABLE 4

EFFECT OF O,P'DDD AND PHENOBARBITAL TREATMENT ON 6β-HYDROXYLATION OF TRIAMCINOLONE ACETONIDE (TrA) BY 9000 X G LIVER SUPERNATE FROM ADULT MALE RATS

Experiment	Treatment	6β-OH-TrA formed in 20 min. (nmoles)
1	Control	75
	o,p'DDD	133
2	Control	73
	Phenobarbital	124

Rats were treated with phenobarbital (70 mg/kg, i.p.) or with o,p'DDD (300 mg/kg, orally) for 3 days. On the 4th day, livers were removed and 9000xg supernate was prepared. Each incubation contained 600 nmoles of TrA. Values represent the mean for 6 animals per group. Incubation conditions as previously described (Kupfer et al,1969b).

would affect the TrA-mediated induction of tyrosine transaminase was examined essentially as described in Table 5. Four hours after the administration of TrA to rats which were pretreated with either phenobarbital or o,p'DDD the rats were killed and the activity of liver tyrosine transaminase assayed. Tables 6 and 7 demonstrate that both phenobarbital and o,p'DDD interfere with the TrA-mediated induction of tyrosine transaminase but, as expected, do not interfere with the cortisol mediated induction of tyrosine transaminase. Furthermore, both phenobarbital and o,p'DDD enhanced

TABLE 5

EXAMINATION OF THE BIOLOGICAL EFFECTS OF INDUCERS OF HEPATIC MONOOXYGENASE

1. Administer inducer (insecticide or drug) for 1 - 4 days
2. Eliminate endogenous source of steroidal hormone (adrenalectomy)
3. One day after last dose of inducer, inject hormone
4. Examine the biological response attributed to the action of the hormone (e.g., induction of tyrosine transaminase)

TABLE 6

THE EFFECT OF PHENOBARBITAL ON N-DEMETHYLATING ACTIVITY AND ON THE INDUCTION OF TYROSINE TRANSAMINASE BY TRIAMICINOLONE (Tr) AND TRIAMCINOLONE ACETONIDE (TrA)

Pretreatment with Phenobarbital	Corticoid Injected	Tyrosine Transaminase Activity	N-Demethylating Activity
-	-	88 ± 19	0.126 ± 0.015 [a]
+	-	91 ± 23	0.211 ± 0.018
-	Tr, 5μg	278 ± 39 [a]	0.125 ± 0.002 [a]
+	Tr, 5μg	161 ± 37	0.175 ± 0.008
-	TrA, 3μg	253 ± 10 [a]	0.131 ± 0.004 [a]
+	TrA, 3μg	162 ± 16	0.190 ± 0.030

[a] $P < 0.05$: probability that the two mean values connected by the vertical line are equal. There were 4-5 rats per group.
Phenobarbital treatment as in Table 3.
Liver preparations from the same animals were used in both assays.
Tyrosine transaminase and N-demethylating activities are expressed as μg of p-hydroxyphenylpyruvate and μmoles of p-chloroaniline, respectively, formed by 67 mg of liver tissue during 10 min incubation.

the N-demethylating activity in the above liver preparations, demonstrating that the monooxygenase system was stimulated. By contrast, SKF 525A, a potent inhibitor of the monooxygenase system, given 30 min prior to TrA, was found to enhance the induction of tyrosine transaminase by TrA (Kupfer, 1968).

Our studies and those of others demonstrate that drugs and chlorinated hydrocarbon insecticides can diminish the biological activity of administered steroids, presumably by accelerating the metabolism of these steroids. Further studies are needed to determine whether the activity of <u>endogeneous</u> steroids will also be affected by inducers of the monooxygenase system.

TABLE 7

THE EFFECT OF O,P'DDD ON THE INDUCTION OF TYROSINE TRANSAMINASE BY TRIAMCINOLONE ACETONIDE (TrA) AND HYDROCORTISONE (F)

Pretreatment with o,p'DDD	Corticoid Injected	Tyrosine Transaminase Activity	N-Demethylating Activity
		Experiment 1	
-	-	67 ± 22	0.153 ± 0.005 [a]
+	-	95 ± 12	0.195 ± 0.009
-	TrA, 3μg	231 ± 8 [a]	0.147 ± 0.019 [a]
+	TrA, 3μg	150 ± 11	0.242 ± 0.032
-	F, 200μg	198 ± 10	0.132 ± 0.006 [a]
+	F, 200μg	187 ± 28	0.197 ± 0.020
		Experiment 2	
-	-	69 ± 9	
+	-	98 ± 14	
-	TrA, 5μg	295 ± 27 [a]	
+	TrA, 5μg	181 ± 11	
-	F, 400μg	265 ± 24	
+	F, 400μg	251 ± 25	

[a] $P < 0.05$; probability that the two mean values connected by a vertical line are similar. Five or six rats were used per group. For details see Table 6.

REFERENCES

Balazs, T. and D. Kupfer. 1966. Effect of DDT on the metabolism and production rate of cortisol in the guinea pig. Toxicol. Appl. Pharmacol. 9: 40-43.

Bergenstal, D.M., R. Hertz, M.B. Lipsett and R.H. Moy. 1960. Chemotherapy of adrenocortical cancer with o,p'DDD. Ann. Int. Med. 53: 672-682.

Conney, A.H. 1967. Pharmacological implications of microsomal enzyme induction. Pharmacol Rev. 19: 317-366.

Cueto, C. and J.H.U. Brown. 1958. Biological studies on an adreno-corticolytic agent and the isolation of the active components. Endocrinology 62: 334-339.

Fahim, M.S., J. Ishaq, D.G. Hall and R.L. Russell. 1970. Effect of DDT on uterine maintenance in rats. Fed. Proc. 29: 781.

Gerboth, G. and U. Schwabe. 1964. Einfluss von gewebsgespeichertem DDT auf die Wirkung von Pharmaka. Arch. exp. Path. u. Pharmak. 246: 469-483.

Ghazal, A., W. Koransky, J. Portig, H.W. Vohland and I. Klempau. 1964. Beschleunigung von Entgiftungsreaktionen durch verschiedene Insecticide. Arch. exp. Path. u. Pharmak. 249: 1-10.

Gnosspelius, Y., H. Thor and S. Orrenius. 1969/1970. A comparative study on the effects of phenobarbital and 3,4-benzpyrene on the hydroxylating enzyme system of rat liver microsomes. Chem.-Biol. Interactions 1: 125-137.

Greim, H., H. Remmer and J.B. Schenkman. 1967. Die Induktion mikrosomaler Enzyme der Rattenleber. Arch. Pharmakol. u. exp. Pathol. 257: 278-279.

Hart, L.G. and J.R. Fouts. 1963. Effects of acute and chronic DDT administration on hepatic microsomal drug metabolism in the rat. Proc. Soc. Exptl. Biol. Med. 114: 388-392.

Hart, L.G., R.W. Shultice and J.R. Fouts. 1963. Stimulatory effects of chlordane on hepatic microsomal drug metabolism in the rat. Toxicol. Appl. Pharmacol. 5: 371-386.

Kuntzman, R. 1969. Drugs and enzyme induction. Ann. Rev. Pharmacol. 9: 21-36.

Kuntzman, R., M. Jacobson, K. Schneidman and A.H. Conney. 1964. Similarities between oxidative drug-metabolizing enzymes and steroid hydroxylases in liver microsomes. J. Pharm. Exptl. Therap. 146: 280-285.

Kuntzman, R., D. Lawrence and A.H. Conney. 1965. Michaelis constants for the hydroxylation of steroid hormones and drugs by rat liver microsomes. Mol. Pharmacol. 1: 163-167.

Kupfer, D. 1968. Alteration in the magnitude of induction of tyrosine transaminase by glucocorticoids. The effects of phenobarbital, o,p'DDD and SKF 525A. Arch. Biochem. Biophys. 127: 200-206.

Kupfer, D. 1969. Influence of chlorinated hydrocarbons and organophosphate insecticides on metabolism of steroids. Ann. N.Y. Acad. Sci. 160: 244-253.

Kupfer, D., T. Balazs and D.A. Buyske. 1964. Stimulation by o,p'DDD of cortisol metabolism in the guinea pig. Life Sci., 3: 959-964.

Kupfer, D. and L.L. Bruggeman. 1966. Determination of enzymic demethylation of p-chloro-N-methylaniline. Assay of aniline and p-chloroaniline. Anal. Biochem. 17: 502-512.

Kupfer, D., L.L. Bruggeman and T. Munsell. 1969a. Studies on the occurrence of N-demethylase activity in adrenal and hepatic preparations from guinea pigs and rats. The effect of various substances on the magnitude of this activity. Arch. Biochem. Biophys. 129: 189-195.

Kupfer, D. and S. Orrenius. 1970. Interaction of drugs, steroids and fatty acids with liver-microsomal cytochrome P-450. Eur. J. Biochem. 14: 317-322.

Kupfer, D. and R. Partridge. 1970. 6β-Hydroxylation of Triamcinolone Acetonide by a hepatic enzyme system. The effect of phenobarbital and 1-Benzyl-2-thio-5,6-dihydrouracil. Arch. Biochem. 140: 23-28.

Kupfer, D., R. Partridge and T. Munsell Jones. 1969b. Alteration in the magnitude of induction of tyrosine transaminase by glucocorticoids. II. Effects of phenobarbital, o,p'DDD and SKF 525A on metabolism of triamcinolone acetonide. Arch. Biochem. Biophys. 131: 57-66.

Levin, W., R.M. Welch and A.H. Conney. 1969. Inhibitory effect of phenobarbital pretreatment on the androgen-induced increase in seminal vesicle weight in the rat. Steroids 13: 155-161.

Lin, E.C.C. and W.E. Knox. 1957. Adaptation of the rat liver tyrosine-α-ketoglutarate transaminase. Biochim. Biophys. Acta 26: 85-88.

Sereni, F., F.T. Kenney and N. Kretchmer. 1959. Factors influencing the development of tyrosine-α-ketoglutarate transaminase activity in rat liver. J. Biol. Chem. 234: 609-612.

Tephly, T.R. and G.J. Mannering. 1968. Inhibition of drug metabolism. V. Inhibition of drug metabolism by steroids. Mol. Pharmacol. 4: 10-14.

SUBJECT INDEX